Solar System Gr. 5-8

Best Value Books

Y0-DEX-729

Table of Contents

Astronomers
Pythagoras 1
Eudoxus of Cnidus 2
Aristotle 3
Eratosthenes 4
Hipparchus 5
Ptolemy 6
Copernicus 7
Brahe 8
Galileo 9
Kepler 10
Newton 11
Herschel 12
Match Facts 1 13
Match Facts 2 14
Acrostic 15
Astronomer Chart 16
Crossword puzzle 17
Word Search 18
Match the Astronomer 19
Astronomer Time Line 20

Planets
Mercury 21
Venus 22
Earth 23
Satellite of Earth 24
Mars 25
Jupiter 26
Saturn 27
Uranus 28
Neptune 29
Pluto 30
Comparing Planets 1 31
Crossword Puzzle 32
Word Search 33
True or False? 34
Who am I? 35

Rank the Planets 36
Label the Map 37
Plan a Vacation 38
Match Planet and Fact ... 39

Stars
What is a Star? 40
Star Color 41
Star Size 42
Life Span 43
True or False? 44
Our Sun 45
Solar Sunspot 46
Solar Flares 47
Solar Prominences 48
Solar Eclipse 49
True or False? 50
Acrostic Puzzle 51
Solar Scramble 52
Milky Way 53
Novas and Supernovas ..54
Black Holes 55
Number Matching 56
True or False? 57
Crossword 58
Word Search 59

Celestial Sights
Constellations 60
Ursa Major 61
Comets 62
Parts of a Comet 63
Halley's Comet 64
Asteroids 65
Meteors 66
Meteor Showers 67
True or False? 68
Who Am I? 69
Acrostic 70

Word Search 71
Space Exploration
History of Rockets 72
Space Race 73
The First Astronauts 74
Early Astronauts 75
Space Time Line 76
Who Am I? 77
Space Shuttle 78
Crew Cabin 79
Spacelab 80
Space Stations 81
Lift Off! 82
Microgravity 83
Eating in Space 84
Supper Is On! 85
Space Age Food 86
Clothing in Space 87
Keeping Space Clean .. 88
Re-Entry 89
Astronaut Acronyms 90
Space Matching 91
Crossword 92
Word Search 93
True or False? 94
Who am I? 95
Acrostic 96
Crew Patch 97
Alien: Friend or Foe? 98

Review
Recall the Facts 99
Recall the Facts 100
Recall the Facts 101
Recall the Facts 102
Recall the Facts 103

The student pages in this book have been specially prepared for reproduction on any standard copying machine.

Kelley Wingate products are available at fine educational supply stores throughout the U. S. and Canada.

Permission is hereby granted to the purchaser of this book to reproduce student pages for classroom use only. Reproduction for commercial resale or for an entire school or school system is strictly prohibited. No part of this book may be reproduced for storage in a retrieval system, or transmitted in any form, by any means (electronic, recording, mechanical, or otherwise) without the prior written permission of the publisher.

Solar System Gr 5-8 **CD-3728**   Printed in the United States Of America   ISBN 0-88724-446-7

About the book...

This book is just one in our Best Value™ series of reproducible, skill oriented activity books. Each book is developmentally appropriate and contains over 100 pages packed with educationally sound classroom-tested activities. Each book also contains free skill cards and resource pages filled with extended activity ideas.

The activities in this book have been developed to help students master the basic skills necessary to comprehend basic science facts about space. The activities have been sequenced to help insure successful completion of the assigned tasks, thus building positive self-esteem as well as the self-confidence students need to meet academic and social challenges.

The activities may be used by themselves, as supplemental activities, or as enrichment material for the science program.

Developed by teachers and tested by students, we never lost sight of the fact that if students don't stay motivated and involved, they will never truly grasp the skills being taught on a cognitive level.

About the author...

Barbara (Suzy) Notarianni, an elementary teacher for over twenty years, has taught in both public and private schools in three states. She has taught the solar system to fourth graders for the past six years, incorporating a hands-on approach to the learning process. Suzy believes that learning should be an interesting, fun, and creative process. She emphasizes full student participation and advocates group interaction for the implementation of lessons. Suzy has a Bachelor of Science degree and her Masters degree in Education.

Senior Editors: Patricia Pedigo and Roger De Santi
Production Supervisor: Homer Desrochers
Production: Debra Olier and Bert Cruz

© 1996 Kelley Wingate Publications

Ready-To-Use Ideas and Activities

The activities in this book will help children master the basic skills necessary to become competent learners. Remember, as you read through the activities listed below and as you go through this book, that all children learn at their own rate. Although repetition is important, it is critical that we never lose sight of the fact that it is equally important to build children's self-esteem and self-confidence if we want them to become successful learners.

Flashcard ideas
The back of this book has removable flash cards that will be great for basic skill and enrichment activities. Pull the flash cards out and cut them apart (if you have access to a paper cutter, use that). Following are several ideas for use of the flashcards.

★ Help the student to learn to read the flash cards by flashing a few words at a time.
★ Students may categorize the words into designated groups.
★ The students may be allowed to create "open sorts", using their own ideas to create categories. (Ex: stars, planets, constellations, etc.)
★ Keep a journal of the words and their definitions as they are introduced.
★ Divide the cards into small groups and have children alphabetize them.
★ Have students locate flash card words in books, magazines, and/or newspapers.
★ Have a child select a number of words and write a paragraph that relates the words in a logical manner.
★ Students can copy the flashcard words and their definitions into a journal as they demonstrate recognition and comprehension mastery.
★ Have students illustrate the flash cards.
★ Play "Bingo". Students select the words for their cards. Teacher gives the definition orally and students mark the words on their cards.
★ Use the flashcards as key terms to outline each skill section as it is introduced.
★ Use specific flashcard words as a guide for highlighting important information about any given skill section.

© 1996 Kelley Wingate Publications

Suggested Classroom Activities

1. After presenting the information on stars and constellations, plan a visit to a planetarium.

2. Use the computer for bulletin board space news.

3. Choose a day of each week for "Space News". Share current space events.

4. Make a collage of pictures depicting space news and events.

5. Visit a space museum.

6. Get information about Space Camp from NASA.

7. Find out how the space program has benefitted man. Have students make a list of products developed by the space program that are a common part of our lives (i.e. velcro, microwaves, etc.).

8. Write letters to science fiction authors or screen play writers. Ask them how they create the setting for their space stories. (Is the story based on fact or completely imagined? Why were the authors interested in space? etc.)

9. Plan an exercise program that could be used in space. (The physical education teacher may help with this.)

10. Build a scale model of the shuttle.

11. Have students list songs that contain lyrics about space.

12. Write to NASA and find out the requirements for becoming an astronaut.

13. Chart the similarities and differences between astronauts and cosmonauts.

14. Hunt for micrometeorites by dragging a magnet through a container of rain water or snow. Scrape the magnet against a microscope slide and examine it to see if it picked up any micrometeorites.

15. Have groups of students become "advertising agencies" that promote travel to a particular planet, constellation, or other feature of space. See if they can "sell" their trip package to other students.

Name _____   Skill: Theory

Pythagoras

Pythagoras was one of the earliest known astronomers. He was born in Greece about 560 B.C. He and a group of other Greeks moved to Croton in southern Italy and set up a school for philosophical and religious ideas. Pythagoras loved mathematics and science, which he applied to the study of the stars and planets. His group believed that the earth revolved around a massive fire contained deep within its core. They also believed that the sun and other planets revolved around this fire.

1. Draw a cross section of the Earth as the Pythagorians thought it might look.

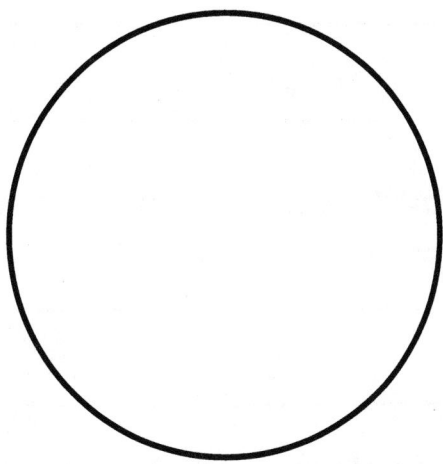

2. What was the contribution that Pythagoras made to our understanding of the solar system?

_____

_____

_____

3. Pretend that we have no information about the solar system. Think of two ways you might learn about the stars and planets. List your methods here.

_____

_____

_____

© 1996 Kelley Wingate Publications   1   CD-3728

Name _____   Skill: Theory

Eudoxus of Cnidus

    Eudoxus of Cnidus was born in Greece around 408 B.C. He was a scholar interested in the motions of celestial bodies (stars, planets, and moons). Eudoxus was the first person to develop the theory that the stars and planets are spheres that move in concentric circles (evenly spaced within each other) around the earth. This theory is called geocentric, or an earth centered system. This theory helped explain why the sun and stars move across the sky from east to west.

1. What did Eudoxus contribute to our understanding of the solar system?

_____

_____

2. Look up Eudoxus in the encyclopedia. What other subjects did he study?

_____

_____

3. Draw a simple diagram of the solar system as Eudoxus believed it worked.

4. "Geo" means earth and "Centric" means centered. Why is the word geocentric a good name for the theory developed by Eudoxus?

_____

_____

© 1996 Kelley Wingate Publications   CD-3728

Name _____   Skill: Theory

Aristotle

    Aristotle, a famous Greek scientist, was born about 384 B.C. When he was seventeen he went to Athens to study under Plato, a famous philosopher and one of the greatest thinkers of all time. Aristotle spent twenty years a Plato's school. During that time he began to observe nature, record what he found, and form theories about how nature worked. He organized information and explained many ideas in studies such as biology, ethics, politics, and astronomy. Aristotle agreed with the geocentric theory of Pythagoras and Eudoxus, believing that the planets revolve around the earth. However, he did not believe that the planets move in large circles. His observations led him to believe that the planets move in elliptical, or oval-shaped orbits. Aristotle was respected as a thinker and so his theory was accepted as fact. The world continued to believe that the earth was the center of the universe for several hundred years until other scientists decided to question this theory.

1. The world believed Aristotle's astrology findings for many years. What are two of those findings presented in this paragraph?

_____

_____

_____

2. Which of Aristotle's findings do we still believe today? Which "fact" has proven to be false?

   True : _____

   False: _____

3. Make your own elliptical orbit. Place two push pins or thumbtacks about 10 centimeters apart on a piece of cardboard. Loop a piece of string (about 20 cm long) around the tacks and tie it tightly. Put your pencil inside the loop and, pulling tightly against the string, trace the path of the string. As the string gives somewhat, the pattern will widen in the center, creating a type of elliptical orbit on paper.

© 1996 Kelley Wingate Publications

Name _____   Skill: Theory

Eratosthenes

Eratosthenes was born about 276 B.C in Cyrene, Greece and became the head librarian in the city of Alexandria, Egypt. Eratosthenes made a major contribution to astronomy by calculating the circumference of the earth. At noon during the summer solstice the sun reflected at the bottom of a well in Syene (modern Aswan). Eratosthenes concluded that this city must be in a direct line between the sun and the center of the earth. He calculated the angle of the sun's shadow in Alexandria at exactly noon on the same day. He found that Alexandria was south of Syene by 1/50 of a circle. He measured the distance between the two cities and multiplied that by 50. Eratosthenes calculated the circumference of the earth to be about 39,000 km (or 24,000 miles). He was very close to the actual circumference as we know it today.

1. Use a world atlas to find the actual circumference of the earth. Record the measurement here.

   _____ kilometers

   _____ miles

2. Calculate the difference between the actual circumference and the measurement Eratosthenes found.

   _____ kilometers

   _____ miles

3. What led Eratosthenes to the conclusion that Syene was in a direct line between the sun and the center of the earth?

   _____
   _____
   _____
   _____

© 1996 Kelley Wingate Publications   4   CD-3728

Name _____ Skill: Theory

Hipparchus

Hipparchus was born in Nicaea (now Iznik, Turkey) about 190 B.C. He was one of the greatest early astronomers and is known as the Father of Systematic Astronomy. Hipparchus believed that the earth was the center of the universe and all other celestial bodies moved around it in perfect circles. He catalogued over 850 stars and kept a record of their small but important yearly movements in our sky. Hipparchus found better ways of determining the distances and diameters of the Sun and Moon. He was one of the first people to use mathematics in determining longitude and latitude of positions on earth.

1. **Look in an encyclopedia to find the distance from the Earth to the Sun and the Moon. Find the diameters of the Sun and Moon.**

   Earth to Sun   _____ km.   _____ mi.

   Earth to Moon   _____ km.   _____ mi.

   Diameter of Sun _____ km.   _____ mi.

   Diameter of Moon _____ km.   _____ mi.

2. **Compare the solar system theory of each of these three men.**

EUDOXUS	ARISTOTLE	HIPPARCHUS
Center of Universe:	Center of Universe:	Center of Universe:
How planets move:	How planets move:	How planets move:

© 1996 Kelley Wingate Publications   CD-3728

Name _____ Skill: Theory

Ptolemy

Ptolemy was born in Greece about 100 A.D. He made many contributions to geography, mathematics, and astronomy. He spent fourteen years in Alexandria, Egypt observing the stars and planets. Ptolemy wrote <u>Almagest</u>, a book that gave a detailed theory involving motion the of the Sun, Moon, and planets. He based his theory on the beliefs of Hipparchus (all celestial bodies revolve around the earth) and Aristotle (planets move in elliptical orbits). Ptolemy added the idea that planets move in epicycles. He believed that the planets were turn around their own centers as they revolve around the earth. His theory was not much improved upon for more than 1000 years.

1. What was the major contribution that Ptolemy gave to the science of astronomy?

_____

_____

_____

_____

2. Ptolemy studied the science of cosmology. What is cosmology? Use an encyclopedia or dictionary to find the answer.

_____

_____

_____

3. Ptolemy later wrote a book about the earth and called it <u>Geography</u> . What did he attempt to do in this book? Use an encyclopedia to find the answer.

_____

_____

_____

© 1996 Kelley Wingate Publications   6   CD-3728

Name _____   Skill: Theory

Copernicus

Nicalous Copernicus was born in Poland on February 19, 1473. He studied mathematics, astronomy, medicine, and law. He practiced medicine and law for his church for many years, but he never lost his love for astronomy. In 1514 at the age of 40, Copernicus wrote a paper challenging the geocentric theory. He agreed with Hipparchus that the planets revolve in perfect circles. He also agreed with Ptolemy that the planets have epicycles, spinning in smaller orbits as they make the larger revolution. However, Copernicus introduced the idea that the earth and planets revolve around the sun. This new idea became known as the heliocentric (sun centered) theory.

1. What was Copernicus' major contribution to the study of astronomy?

_____

_____

_____

2. In what ways did Copernicus agree with Hipparchus and Ptolemy?

_____

_____

_____

_____

_____

3. Look up geocentric and heliocentric in the dictionary. What does each word mean?

GEOCENTRIC _____

_____

HELIOCENTRIC _____

_____

© 1996 Kelley Wingate Publications   CD-3728

Name _____   Skill: Theory

Brahe

Tycho Brahe was born December 14, 1546 in Denmark. He studied law, but became interested in astronomy when he observed a solar eclipse in 1560. He read Ptolemy's <u>Almagest</u> and went on to study science in several universities. In 1572 Brahe discovered a nova (a star that becomes very bright then fades in a few months or years) and five comets that were beyond the Moon's orbit. His findings did not agree with either Ptolemy or Copernicus so Brahe developed his own theory called the Tychonic system. This system suggests that the Sun and Moon revolve around the Earth and all other planets revolve around the Sun. Although his theory is not accurate, Brahe did contribute a great deal to the science of astronomy. He developed new and more accurate instruments for measuring the movement of stars and planets. He also compiled a large collection of observations that became useful to his student and future astronomer, Kepler.

1. What was the major contribution Brahe gave to astronomy?

_____

_____

_____

2. Explain in your own words what a nova and a comet are.

    NOVA _____

_____

_____

    COMET _____

_____

_____

3. On the back of this paper, draw a diagram of how Brahe thought the universe must look.

© 1996 Kelley Wingate Publications

Name _____ Skill: Theory

Galileo

Galileo Galilei was born in Italy on February 15, 1564. As a young boy Galileo hated science because he did not believe what was taught. He felt theories needed to be proved before they could be believed. Galileo discovered that mathematics could provide a way to test (and prove or disprove) scientific theories. In 1604 Galileo became interested in astronomy when a Supernova appeared. In 1609 he was introduced to a Dutch toy that made far away things look closer - the first telescope. Galileo designed and built his own much larger telescope and turned it toward the sky. He observed the Moon and several "planets" circling Jupiter (he later identified these as moons). Galileo noticed that Venus had phases just like the moon. This observation gave him the first proof that the planets did indeed revolve around the Sun as Copernicus had stated.

1. Why didn't Galileo care much for science? What changed his mind?
_____
_____
_____
_____

2. In your opinion, which was Galileo's greater discovery - the use of the telescope or proof that the planets circled the Sun? Justify your answer with facts.
_____
_____
_____
_____

3. Did Galileo's findings support a geocentric or heliocentric theory?
_____
_____
_____
_____

© 1996 Kelley Wingate Publications   CD-3728

Name _____ Skill: Theory

Kepler

Johannes Kepler was born on December 27, 1571 in Germany. He studied astronomy under Tycho Brahe and supported most of the theories developed by Copernicus. Kepler observed the orbit of planets and was the first to accurately determine their position and how they orbited the sun. Kepler developed three laws regarding planetary motion. The first law states that the sun is the center of the universe and that the planets revolve around it in elliptical orbits. The second law states that planets move faster when their orbits bring them closer to the sun. The third law is a mathematical formula used to determine the distance of a planet from the sun.

1. Kepler proved the planets orbited the sun in elliptical orbits. How did this agree with the teachings of Copernicus? How did this disagree with Copernicus?

   AGREE _____
   _____
   _____

   DISAGREE _____
   _____
   _____

2. Kepler wrote six books during his lifetime. Look in the encyclopedia and find the names of two of his books and the date they were published.

   _____
   _____

3. What was the major contribution that Kepler made to our understanding of the solar system?

   _____
   _____

© 1996 Kelley Wingate Publications   CD-3728

Name _____ Skill: Theory

Newton

Sir Isaac Newton was born on January 4, 1643 in England. He is often noted as the greatest scientific genius of all time because he made important contributions to every major area of science known during his time: Mathematics, physics, optics, and astronomy. In the field of astronomy, Newton is best known for his laws of motion and gravity. People already knew that objects dropped would fall to the ground, but they had no idea why that happened. Newton explained that there was a force called gravity that pulled objects toward the center of the earth. Newton developed the theory of gravitational pull. Gravity is a force that pulls objects toward the center of an area. The Earth has a gravitational pull that attracts even the moon! The sun also has gravity that tries to pull the Earth and other planets toward its center. Newton explained a law of motion called centrifugal force that stops the planets from being pulled into the sun. As an object revolves in a circular path, centrifugal force causes the object to pull away from the center of the orbit. A balance between gravity and centrifugal force holds the planets in a steady orbit around the sun!

1. What important contribution did Newton make to the science of astronomy?

_____
_____
_____

2. According to Newton, why does a planet orbit around the sun?

_____
_____
_____

3. Newton wrote a book that is considered to be the greatest scientific book ever written. What is the name of that book? Use an encyclopedia to help you find the answer.

_____

© 1996 Kelley Wingate Publications   CD-3728

Name _____ Skill: Theory

Herschel

Sir William Herschel was born in Germany on November 15, 1738. He was a music teacher but spent his spare time studying astronomy and building telescopes. He constructed the largest reflecting telescopes of his time using them to observe the six known planets: Mars, Mercury, Earth, Venus, Jupiter, and Saturn. In 1781 Herschel looked beyond the known solar system and made an amazing discovery - a new planet! He named this seventh planet "Georgium Sidus" after England's King George III. In 1782 Herschel constructed an even larger telescope and found Georgium Sidus has two moons. He also discovered two new moons of Saturn. The name Herschel gave his planet never really stuck. Astronomers preferred to call the planet "Herschel", but by the mid 1800's it was was commonly known by a name taken from mythology.

1. What was Sir William Herschel's contribution to the science of astronomy?

_____
_____
_____
_____

2. Using the encyclopedia, look up Sir William Herschel and find the name that we commonly use for the planet he called Georgium Sidus.

_____
_____
_____
_____

3. Pretend that you have just discovered a new planet in our solar system. What would you name the planet? Explain why you chose that name.

_____
_____
_____
_____

© 1996 Kelley Wingate Publications   CD-3728

Name _____   Skill: Theory

Matching Astronomers

Match the astronomer to the fact about him.

ASTRONOMER

_____ Aristotle

_____ Brahe

_____ Copernicus

_____ Eratosthenes

_____ Eudoxus

_____ Galileo

_____ Hipparchus

_____ Herschel

_____ Kepler

_____ Newton

_____ Ptolemy

_____ Pythagoras

FACT

A. He discovered the seventh planet in 1781.

B. He wrote the Almagest

C. He developed the theory that planets are spheres and move in concentric circles around the Earth.

D. He introduced the idea of elliptical orbits in geocentric theory.

E. He developed the Tychonic system and was the first to discover a nova.

F. He used his telescope to prove the heliocentric theory.

G. He explained gravity and centrifugal force.

H. He used math to determine longitudes and latitudes of the Earth.

I. He believed the planets revolve around a fire at the core of the Earth.

J. He wrote three famous laws about the movement of planets.

K. He measured the distance around the Earth.

L. The first astronomer to name the sun as the center of the solar system.

© 1996 Kelley Wingate Publications   13   CD-3728

Name _____ Skill: Theory

Matching Astronomers

Match the astronomer to the fact about him.

ASTRONOMER

_____ **Aristotle**

_____ **Brahe**

_____ **Copernicus**

_____ **Eratosthenes**

_____ **Eudoxus**

_____ **Galileo**

_____ **Hipparchus**

_____ **Herschel**

_____ **Kepler**

_____ **Newton**

_____ **Ptolemy**

_____ **Pythagoras**

FACT

A. He challenged the geocentric theory in 1514.

B. He discovered Georgium Sidus.

C. He developed the theory that planets move in concentric circles around the Earth.

D. He studied under Plato for twenty years.

E. He was called the Father of Systematic Astronomy.

F. He set up a school for philosophy and religion in Croton, Italy.

G. He proved heliocentric theory by discovering the phases of Venus.

H. He is called the greatest scientific genius of all time.

I. He believed that the planets move in epicycles.

J. He was the head librarian in Alexandria, Egypt.

K. He was the first to discover a nova (1572).

L. He studied under Tycho Brahe.

© 1996 Kelley Wingate Publications   14   CD-3728

Name _____   Skill: Theory

Read the clues and figure out the name of the astronomer. Write his name in the spaces provided. (Clue number one goes first and begins with an A.)

```
                                    A _ _ _ _ _ _ _ _
                              _ _ _ S _ _ _ _
                              _ _ _ T _ _
                        _ _ _ _ _ _ _ R _ _
                        _ _ _ _ _ _ _ O
            _ _ _ _ _ _ _ _ _ _ N _ _
                        _ _ _ _ _ _ _ O _ _ _ _ _ _ _
                              _ _ _ _ _ M _
                              _ _ _ _ E _
                                    _ R _ _ _
```

1. He studied under another astronomer named Plato.

2. He named a new planet "Georgium Sidus".

3. He is noted as the greatest scientific genius of all time.

4. He believed that there was a huge fire inside the Earth.

5. He turned a Dutch toy into an instrument to view the sky more clearly.

6. He measured the sun's angle between two cities to find the circumference of the Earth.

7. He introduced the heliocentric theory.

8. He added the theory of epicycles to the study of astronomy.

9. He developed a mathematical formula to determine the distance between the planets and the sun.

10. He believed the Sun and Moon revolve around the Earth, but all other planets revolve around the Sun.

© 1996 Kelley Wingate Publications   CD-3728

Name _____   Skill: Theory

Astronomer Chart

Look at the list of astronomers below the chart. List them under the proper heading according to which system they believed in - geocentric or heliocentric.

GEOCENTRIC	HELIOCENTRIC

Copernicus	**Aristotle**	**Hipparchus**
Galileo	**Ptolemy**	**Pythagoras**
Eudoxus	**Kepler**	**Newton**
	Herschel	

© 1996 Kelley Wingate Publications — CD-3728

Name _____ Skill: Theory

ASTRONOMER CROSSWORD

ACROSS

1. He gave us the laws of planetary motion.
3. He measured the circumference of the Earth.
5. He believed planets moved in elliptical orbits.
6. This man discovered Uranus.
9. He introduced the theory of epicycles.
11. This man was the first to believe in a heliocentric theory.
12. He gave us the laws of gravity.

DOWN

2. He was born about 560 B.C.
4. Using his telescope, this man proved heliocentric theory.
7. He introduced planet orbits as concentric circles.
8. He developed lines of longitude and latitude for the Earth.
10. He developed the Tychonic System to explain our solar system.

© 1996 Kelley Wingate Publications
CD-3728

Name _____ Skill: Theory

ASTRONOMERS

Circle the words listed as you find them in the word search.
They may go in any direction.

F	P	R	C	A	S	T	R	O	N	O	M	Y	S	P
J	G	L	E	I	S	U	X	O	D	U	E	O	E	B
G	E	A	A	L	R	I	K	W	F	D	L	R	R	E
T	O	R	H	N	P	T	E	F	X	A	A	A	O	P
E	C	I	E	O	E	E	N	Q	R	T	H	B	E	O
S	E	S	R	D	B	T	K	E	O	E	N	X	L	C
U	N	T	S	P	R	B	G	S	C	N	I	N	I	S
C	T	O	C	V	N	F	T	E	E	O	Z	E	L	E
I	R	T	H	S	F	H	L	W	N	T	I	L	A	L
N	I	L	E	Z	E	C	T	K	I	O	I	L	G	E
R	C	E	L	N	Y	O	A	T	M	N	V	B	E	T
E	J	H	E	C	N	V	K	H	D	N	I	A	R	H
P	V	S	I	P	Y	T	H	A	G	O	R	A	S	O
O	Z	P	H	I	P	P	A	R	C	H	U	S	S	R
C	E	E	L	L	I	P	T	I	C	A	L	K	U	S

ARISTOTLE   ASTRONOMY   BRAHE
COPERNICUS   ELLIPTICAL   EPICYCLE
ERATOSTHENES   EUDOXUS   GALILEO
GEOCENTRIC   HELIOCENTRIC   HERSCHEL
HIPPARCHUS   KEPLER   NEWTON
NOVA   ORBIT   PLANET
PYTHAGORAS   SOLAR   TELESCOPE

© 1996 Kelley Wingate Publications   CD-3728

Name _____ Skill: Theory

Matching Systems

Match the astronomer to the solar system he believed in.

| **Aristotle** | **Copernicus** | **Eudoxus** | **Pythagoras** |

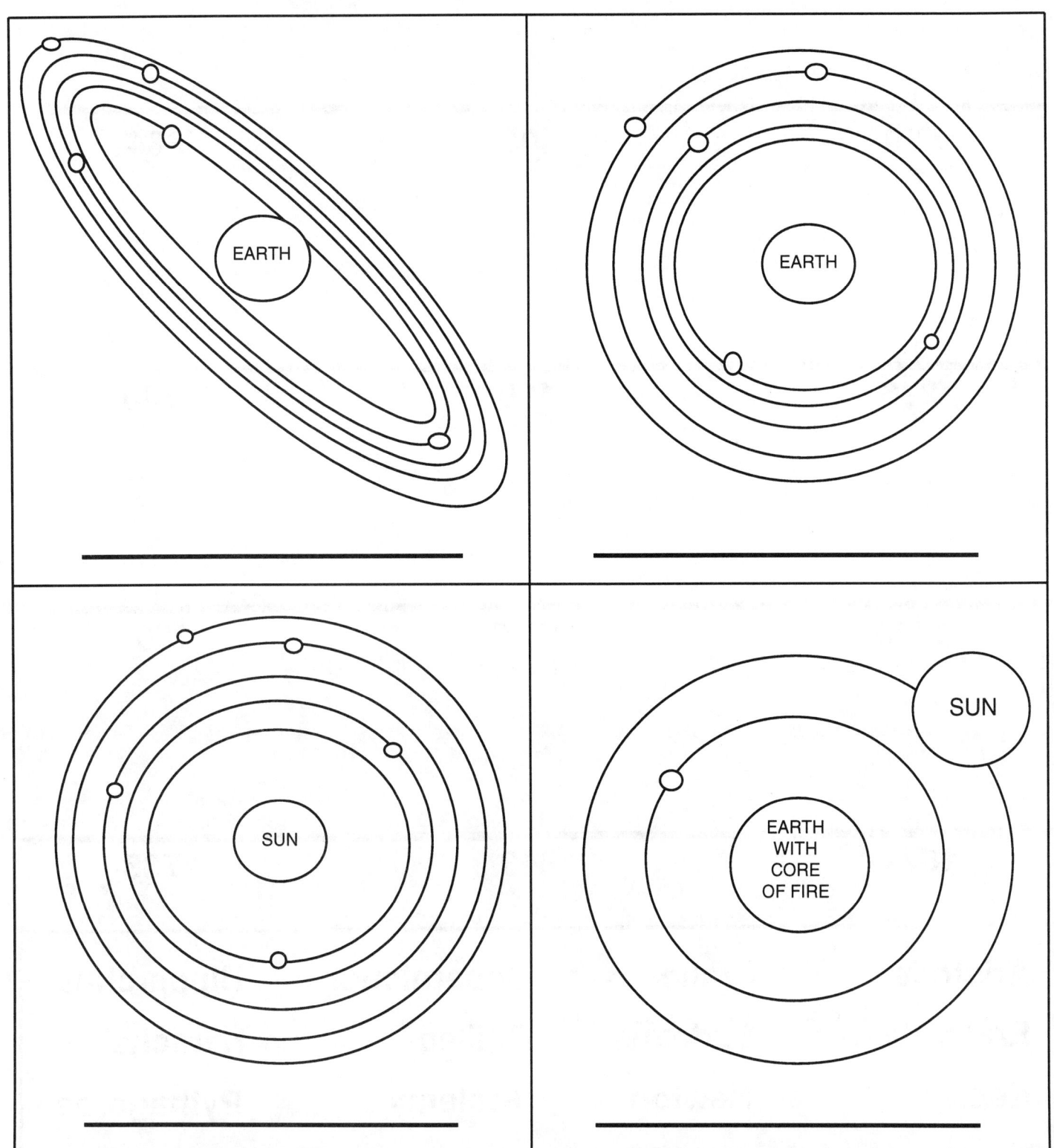

Name _____ Skill: Space Exploration

Astronomer Time Line

Match each astronomer with the year of his birth. Write his name over the correct date on this time line.

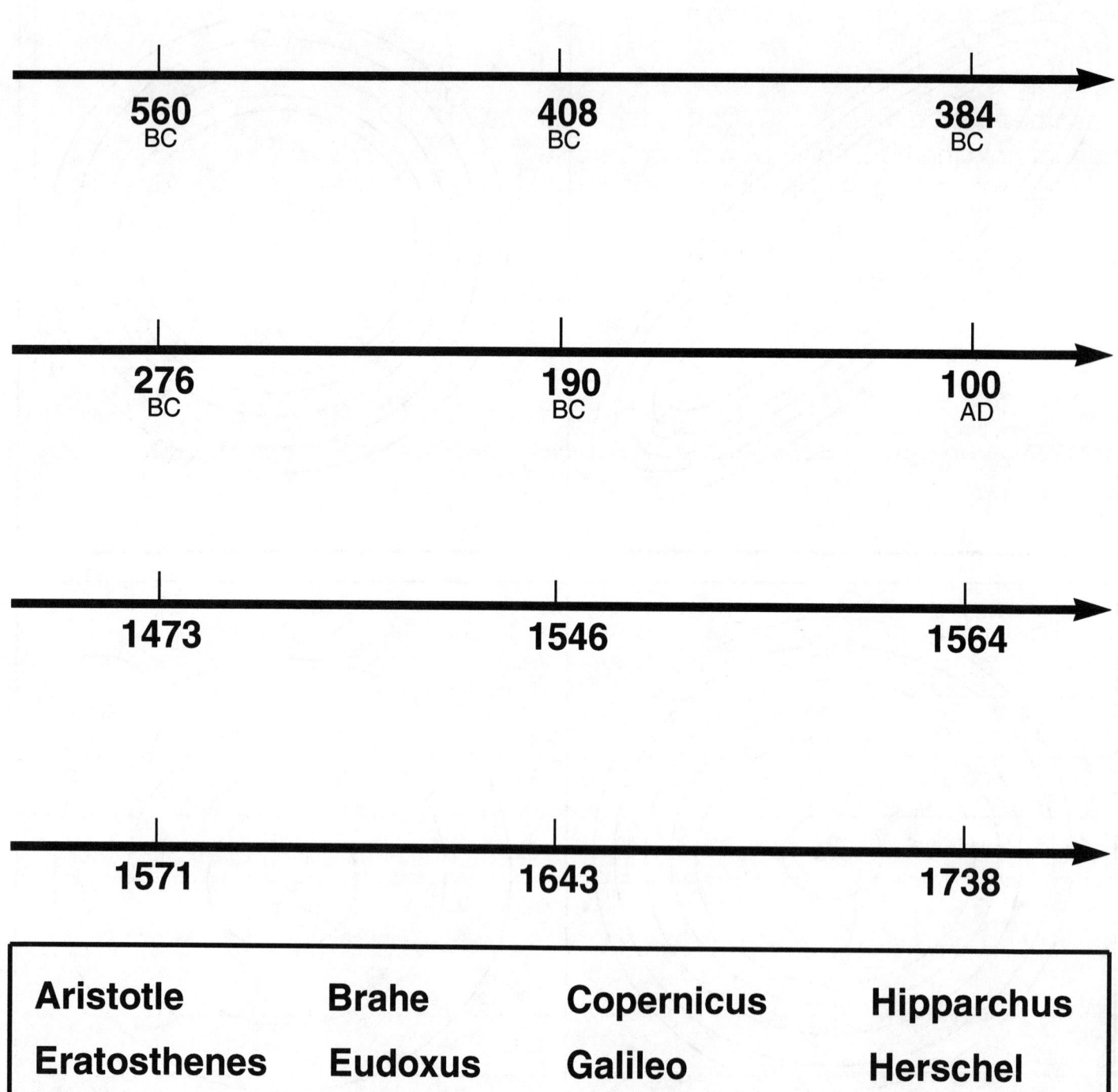

Aristotle	Brahe	Copernicus	Hipparchus
Eratosthenes	Eudoxus	Galileo	Herschel
Kepler	Newton	Ptolemy	Pythagoras

© 1996 Kelley Wingate Publications

Name _____   Skill: Planets

Mercury

Mercury is the closest planet to our Sun. It is a small planet, only a littler larger than our moon. Mercury is one of the five planets that can be seen with the naked eye, but it is the most difficult to see. Mercury rotates slowly. One day on Mercury is equal to 59 Earth days! The atmosphere on Mercury is very thin, allowing a lot of heat to be absorbed during the day. At night the atmosphere cannot hold the heat so it becomes extremely cold very quickly. Mercury is covered with craters made by meteors crashing into it. One large area of craters is called Caloris Basin. Mercury's orbit is unusual in that it is more elliptical (oval shaped) than most other planets.

Fact Box
Distance from Sun: 57.8 million kilometers (36 million miles)
Rotation: 59 Earth days
Revolution: 88 Earth days
Diameter: 4,880 kilometers (3,030 miles)
Gravity: about 1/3 of the Earth's gravity
Orbital Speed: 48 kilometers per second (30 miles per second)
Atmosphere: extremely thin
Temperature: 350° C (660° F) day -170° C (-270° F) at night
Rings: none
Satellites: none
Travel time from Earth:
Jet: 10 years, 8 months
Rocket: 3 months
Light Years: 5 minutes
Named for: Roman god, Mercury (Greek god, Apollo)

1. If you were staying on Mercury for one week, how long would that be in Earth days?

_____

2. Why is Mercury's temperature so hot during the day then so cold at night?

_____

_____

_____

_____

3. What does the surface of Mercury look like? Describe it in your own words.

_____

_____

_____

© 1996 Kelley Wingate Publications   CD-3728

Name _____   Skill: Planets

Venus

Venus is sometimes called the sister planet to Earth because they are so close in size. However, Venus rotates the opposite of the Earth, with the sun rising in the west and setting in the east. The atmosphere of Venus is very thick. If you were to stand on that planet the air would feel like 3,000 feet of water crushing you! Venus has a thick cloud layer that helps reflect the sun so it looks very bright in the night sky. Those clouds also hold the sun's heat next to the surface of the planet, making it extremely hot. The surface of Venus is similar to Earth. It has mountains, valleys, plains, continents, and even areas where oceans could be (if there were any water). Venus also has volcanoes, some of which are many miles wide.

Fact Box
Distance from Sun: 108 million kilometers (67 million miles)
Rotation: 243 Earth days
Revolution: 225 Earth days
Diameter: 12,100 kilometers (7,520 miles)
Gravity: about 9/10 of the Earth's gravity
Orbital Speed: 35 kilometers per second (22 miles per second)
Atmosphere: 98% carbon dioxide
Temperature: 400° C (800° F)
Rings: none
Satellites: none
Travel time from Earth:
Jet: 5 years, 5 months
Rocket: 1.5 months
Light Years: 2.5 minutes
Named for: Roman goddess Venus (Greek goddess Aphrodite)

1. Why does Venus shine so brightly in the night sky?

_____

_____

_____

2. Name six ways in which Venus and the Earth are alike:

   1. _____   2. _____

   3. _____   4. _____

   5. _____   6. _____

3. If you visited Venus for one week, how many Earth days would that be? How many Earth years?

   Seven Venus days equal _____ Earth days.

   Seven Venus days equal _____ Earth years.

© 1996 Kelley Wingate Publications   22   CD-3728

Name _____ Skill: Planets

Earth

Earth is the third planet from the Sun and the only known planet to have life. The Earth looks very beautiful from space. It is mostly blue (three-fourths of its surface is covered with water), with brown and green continents and white ice-capped poles. At any given time about half of the Earth is covered with clouds which protect the surface from extreme heat. The clouds also retain, or hold, the heat that is absorbed so that the planet does not become too cold as it rotates away from the sun. The Earth is tilted on its axis, rotating at an angle which allows us to experience seasons. The highest point on this planet is Mount Everest at 5.5 miles above sea level. The lowest point is found in the Dead Sea and is 395 meters (1,296 ft) below sea level.

Fact Box
Distance from Sun: 150 million kilometers (93 million miles)
Rotation: 23 hrs 56 min
Revolution: 365.3 days
Diameter: 12,800 kilometers (7,900 miles)
Gravity:
Orbital Speed: 24 kilometers per second (15 miles per sec.)
Atmosphere: 78% nitrogen, 21% oxygen, 1% other gases
Temperature: varies with location
Rings: none
Satellites: one (the Moon)
Travel time from Earth:
Jet:
Rocket:
Light Years:
Named for: Greek goddess Gaea, mother of the Titans

1. What makes the Earth different from every other planet in our solar system?

_____

_____

2. How do clouds help protect our planet?

_____

_____

_____

3. What feature of the Earth gives it seasons?

_____

_____

_____

© 1996 Kelley Wingate Publications

Name _____   Skill: Satellite

The Moon

The only natural satellite of the planet Earth is the Moon. It is about one fourth the size of the Earth and has no air, food, or water. The gravity of the Moon is too weak to hold an atmosphere although it is strong enough to cause ocean tides on the Earth. The Moon rotates at the same rate it revolves, so the same side always faces the Earth. Humans did not know what was on the dark side of the Moon until 1959 when a Russian spacecraft brought back pictures. In 1969 astronauts walked on the surface of the Moon for the first time. About three fourths of the Moon is covered with mountains and craters formed by crashing meteors. The craters are as large as 1000 km. (620 mi.) across and as small as dots. The rest of the Moon is smooth and dark, commonly called "seas".

Fact Box
Distance from Earth: 385,000 kilometers (239,000 miles)
Rotation: 27 days, 7 hrs, 43 mins
Revolution: 27 days, 7 hrs, 43 mins
Diameter: 3,476 kilometers (2160 miles)
Gravity: 1/6 of Earth
Orbital Speed: 1 kilometer per second (6 miles per second)
Atmosphere: none
Temperature: 1300 C (2660 F) day to -1730 C (-2800 F) at night
Rings: none
Satellites: none
Travel time from Earth:
Jet: 4 months
Rocket: 3 days
Light Years: 1.2 seconds

1. Why does the same side of the Moon always face the Earth?

_____

_____

2. What caused all of the craters on the Moon?

_____

3. List three facts about the Moon's gravity:

  1. _____

  2. _____

  3. _____

4. Why does the Moon get so hot during the day and so cold at night?

_____

© 1996 Kelley Wingate Publications   24   CD-3728

Name _____ Skill: Planets

Mars

Mars is the fourth planet from the Sun. It is a rocky planet that is much smaller and colder than Earth. The atmosphere of Mars is too cool for water to exist, but the two poles are covered with a thin layer of ice. Mars is also called "the Red Planet" because a large amount of iron in the soil gives it a reddish color. An interesting feature of this planet is a gigantic canyon named Valles Marineris that stretches for 4,000 km (2,500 mi.). During the 1700's, astronomers observed swirls and dark patches thought to be clouds and seas. It was believed that Mars had an atmosphere that could support life. Many fictional stories centered on the possibility of "Martians". In recent years several space probes have collected information about Mars that proves no life could exist on the surface of that planet.

Fact Box
Distance from Sun: 227 million kilometers (140 million mi.)
Rotation: 24 hrs 40 mins.
Revolution: 687 Earth days
Diameter: 6,800 kilometers (4,200 miles)
Gravity: about 1/3 of the Earth's gravity
Orbital Speed: 24 kilometers per second (15 miles per second)
Atmosphere: 95% carbon dioxide, 3% nitrogen, 2% argon
Temperature: -53° C (64° F) to -128° C (-199° F)
Rings: none
Satellites: two
Travel time from Earth:
Jet: 8 years, 10 months
Rocket: 2.5 months
Light Years: 4 minutes
Named for: Roman god, Mars (Greek god Ares)

1. Why is Mars often called the Red Planet?

_____

_____

2. What observations led earlier astronomers to believe Mars might support life?

_____

_____

_____

3. How do we know so much about the surface of Mars when no man has ever set foot on the planet?

_____

_____

_____

© 1996 Kelley Wingate Publications CD-3728

Name _____   Skill: Planets

Jupiter

Jupiter, the fifth planet from the Sun, is the first of the four gas planets. Jupiter is so large that 2/3 of all the planets in our solar system could fit inside it! The planet is covered with three thick layers of clouds that extend 30 km (19 mi.) above the surface. There is no solid surface on the planet. Jupiter is composed of gases that become liquids about 1/4 of the way to the core (center). Because it rotates so quickly, Jupiter appears flattened around the center and longer at the poles. An interesting feature of Jupiter is a number of large white spots that space probes have identified as storms. There is also the Great Red Spot which is as large as the Earth. This is a major storm that has continued to rage for over 300 years!

Fact Box

Distance from Sun: 778 million kilometers (480 million miles)
Rotation: 9 hrs. 55 min.
Revolution: 11.9 Earth yrs.
Diameter: 143,200 kilometers (89,000 miles)
Gravity: 2 1/2 times greater than Earth's gravity
Orbital Speed: 13 kilometers per second (8 miles per second)
Atmosphere: hydrogen and helium
Temperature: At the top of the cloud cover -130° C (-202° F)
Rings: 3
Satellites: 16
Travel time from Earth:
Jet: 74 years, 3 months
Rocket: 1 year, 9 months
Light Years: 35 minutes
Named for: Roman god, Jupiter (Greek god, Zeus)

1. Jupiter has sixteen satellites. Name the four largest ones discovered by Galileo. Use an encyclopedia to help find the answer.

   1. _____
   2. _____
   3. _____
   4. _____

2. If you could choose between sleeping a full night on Earth or a full night on Jupiter, which one would you pick? Explain your reasoning. (Hint: Think about the rotation of each planet.)

_____
_____
_____
_____

© 1996 Kelley Wingate Publications   CD-3728

Name _____   Skill: Planets

Saturn

Saturn is composed of gases, like Jupiter, and has no solid surface area. Because of the quick rotation, the planet is flattened somewhat at the poles and wider around the equator. Saturn's atmosphere is different in that it has two wind systems operating in opposite directions. One system blows east to west and the other blows west to east. Saturn has often been considered the most beautiful planet because of its huge ring system. The rings extend almost the same distance as the space between Earth and the Moon! The rings were discovered by Galileo and are composed of rock and ice particles. The particles range in size from tiny pebbles to car-sized boulders. Saturn has more satellites than any other planet.

Fact Box
Distance from Sun: 1.4 billion kilometers (893 million miles)
Rotation: 10 hrs 40 mins
Revolution: 29 1/2 Earth yrs
Diameter: 120,600 kilometers (74,980 miles)
Gravity: 1.07 times Earth's gravity (almost the same)
Orbital Speed: 9.7 kilometers per second (6 miles per second)
Atmosphere: Hydrogen and helium
Temperature: At the top of the cloud cover -176° C(-285° F)
Rings: 8
Satellites: 24 known
Travel time from Earth:
Jet: 150 years, 5 months
Rocket: 3 years, 7 months
Light Years: 1 hr 11 mins
Named for: Roman god, Saturn (Greek god, Cronus)

1. Name the largest moon (satellite) of Saturn. Use the encyclopedia to help find the answer.

_____

_____

2. What is unusual about the wind systems of Saturn?

_____

_____

3. In what year did Galileo discover the rings of Saturn? Find the answer in your encyclopedia.

_____

_____

4. Why is there no solid surface on the planet Saturn?

_____

_____

© 1996 Kelley Wingate Publications   CD-3728

Name _____   Skill: Planets

Uranus

Uranus, the seventh planet from the Sun, is also a gas planet. It was discovered in 1781 by Sir William Herschel. Through a telescope Uranus appears bluish-green in color. A spacecraft, Voyager 2, passed close to this planet in 1986. The pictures that were sent back to Earth showed no geographical features. However, the Voyager 2 did prove that Uranus has eleven rings, ten narrow and one large, composed of particles no smaller than 20 cm (8 in). Uranus has five major satellites that can be spotted through a telescope: Miranda, Ariel, Umbriel, Titania, and Oberon. Space probes have proved that Uranus also has ten much smaller satellites that orbit much closer to the planet. These ten are dark in color and do not reflect as much sunlight as the larger ones.

Fact Box
Distance from Sun: 2.9 billion kilometers (1.8 billion miles)
Rotation: 17.3 hours
Revolution: 84 Earth years
Diameter: 51.100 kilometers (31,750 miles)
Gravity: Almost the same as Earth's (.91 times)
Orbital Speed: 6.8 kilometers per second (4 miles per second)
Atmosphere: 84% hydrogen, 15 % helium, 1% other
Temperature: -214 C (-353 F) at top of cloud cover
Rings: 11
Satellites: 15
Travel time from Earth:
Jet: 318 years, 6 months
Rocket: 7 years, 7 months
Light Years: 2 hrs, 30 mins
Named for: Roman god, Uranus

1. Find out which astronomers discovered the five larger moons of Uranus. Also, record the year in which each moon was discovered.

	Discovered by	Year
Miranda	_____	_____
Ariel and Umbriel	_____	_____
Titania and Oberon	_____	_____

2. Why weren't the ten smaller satellites discovered until a space probe passed close to Uranus?

_____

_____

_____

© 1996 Kelley Wingate Publications   28   CD-3728

Name _____   Skill: Planets

Neptune

Neptune, the eighth planet from the Sun, is the last gas planet. The planet appears blue because of the methane gas in its atmosphere. Neptune was discovered in 1846 due to the work of two astronomers - John Adams and Jean Leverrier. Several dark spots on the surface are believed to be storms. The largest, called the Great Dark Spot, is the size of Earth. A smaller storm called Scooter has a bright interior that is believed to be wispy clouds. Two satellites, Triton and Nereid, were discovered soon after the planet was identified. Triton is most unusual because it has a retrograde (backwards) orbit around Neptune! The space probe Voyager 2 passed about 5,000 km (3,100 mi) above this planet in 1989, revealing six other smaller satellites.

Fact Box
Distance from Sun: 4.5 billion kilometers (2.8 billion mi)
Rotation: 16 hours
Revolution: 165 Earth years
Diameter: 49,500 kilometers (30,750 miles)
Gravity: Almost the same as Earth's gravity (1.15 times)
Orbital Speed: 5.4 kilometers per second (3 miles per second)
Atmosphere: hydrogen, helium, and methane
Temperature: -218 C (-360 F) at cloud cover
Rings: 5
Satellites: 8
Travel time from Earth:
Jet: 513 years, 2 months
Rocket: 12 years, 3 months
Light Years: 4 hrs, 2 mins.
Named for: Roman god, Neptune (Greek god, Poseidon)

1. What is the Great Dark Spot on Neptune?

_____

_____

2. Who discovered Triton and Nereid? In what year?

	Discovered by	Year
Triton	_____	_____
Nereid	_____	_____

3. What makes Neptune look blue?

_____

4. What makes Triton the most unusual satellite in the solar system?

_____

_____

© 1996 Kelley Wingate Publications   CD-3728

Name _____   Skill: Planets

Pluto

Pluto is the smallest and coolest of the planets. Because it is so far from Earth, there is still much to learn about this planet. It is usually the ninth planet, but not always. Pluto has such an elliptical orbit that, when it is closest to the Sun, it travels inside Neptune's orbit. For about 20 years out of its 248 year revolution, Pluto is actually the eighth planet and Neptune is the ninth! Pluto's surface is covered with nitrogen, carbon monoxide, and methane frozen into an unusual snow. If you were to step onto Pluto it would feel something like very soft pudding. The surface is red near the equator and bluer at the poles. Pluto's one moon, called Charon, is nearly the same size as the planet it orbits.

Fact Box
Distance from Sun: 5.9 billion kilometers (3.7 billion mi)
Rotation: 6 days, 9 hours
Revolution: 248 Earth years
Diameter: 2,300 kilometers (1,500 miles)
Gravity: 1/100 that of Earth's gravity
Orbital Speed: 4.7 kilometers per second (2.9 mi. per second)
Atmosphere: nitrogen, carbon monoxide, and methane
Temperature: -230° C (-382° F)
Rings: none
Satellites: 1
Travel time from Earth:
Jet: 690 years, 1 months
Rocket: 16 years, 5 months
Light Years: 5 hrs, 25 mins.
Named for: Roman god, Pluto (God of death)

1. Why haven't scientists learned as much about Pluto as they know about other planets?

_____

_____

2. What is unusual about the orbit of Pluto?

_____

_____

_____

_____

3. Describe the surface of Pluto in your own words:

_____

_____

_____

_____

© 1996 Kelley Wingate Publications   CD-3728

Name _____ Skill: Planets

Comparing Planets
Answer the questions below using the information from the Fact Boxes for each planet.

1. Which planet is farther from the Sun: Mars or Venus? _____
   How many kilometers farther? _____

2. Which planet has a larger diameter: Neptune or Uranus? _____
   How many kilometers larger? _____

3. Which planet has a shorter rotation time: Earth or Mars? _____
   How many hours and minutes less? _____

4. Which planet is closer to the Sun: Venus or Mercury? _____
   How many kilometers closer? _____

5. Which planet has more rings: Saturn or Jupiter? _____
   How many more? _____

6. Which planet has fewer satellites: Earth or Uranus? _____
   How many fewer? _____

7. Which planet has the slower orbital speed: Saturn or Jupiter? _____
   How many kilometers per second slower? _____

8. Which planet has the shorter travel time (by jet) from Earth:
   Saturn or Mercury? _____
   How many years and months less? _____

9. Which planet has the shorter travel time (in light years) from Earth:
   Jupiter or Venus? _____
   How many minutes shorter? _____

10. Which planet has the shorter revolution around the Sun:
    Venus or Mercury? _____
    How many days shorter? _____

© 1996 Kelley Wingate Publications

Name _____ Skill: Planets

PLANET CROSSWORD

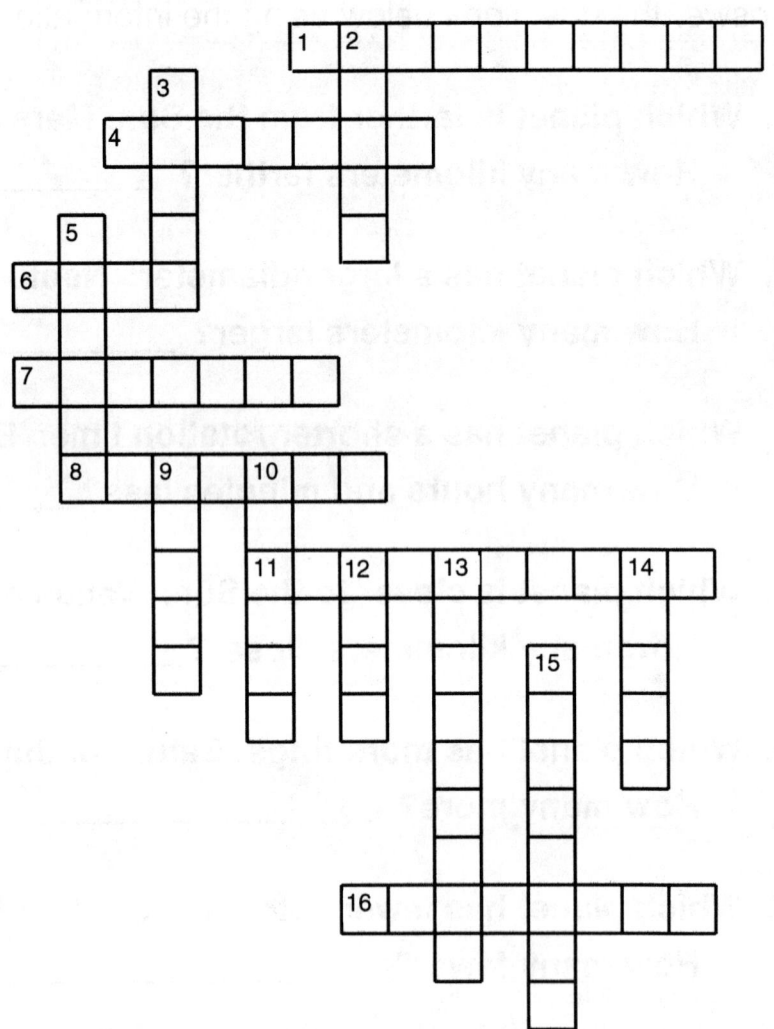

ACROSS

1. The name of an orbit around the Sun.
4. Caloris Basin is found on this planet.
6. The Red Planet.
7. The Great Red Spot is found on this planet.
8. The Great Dark Spot and Scooter are found on this planet.
11. The gases that surround a planet are this.
16. One complete turn of a planet on its axis (one day).

DOWN

2. The third planet from the Sun.
3. The sister planet of Earth.
5. The planet with the most satellites.
9. The smallest and coolest planet.
10. This planet has the moons Miranda, Umbriel, and Ariel.
12. This causes the tides on Earth.
13. A heavenly body that orbits a planet.
14. Rock and ice particles that create a circle around a planet.
15. The distance around the widest part of a planet.

© 1996 Kelley Wingate Publications   32   CD-3728

Name _____   Skill: Theory

PLANETS

Read the clues below and give the name of what it describes.
Circle the names in the word search. They may go in any direction.

P	A	R	O	T	A	T	I	O	N	I
W	L	T	S	A	T	U	R	N	N	N
H	M	U	M	D	W	M	I	O	O	N
G	E	I	T	O	R	U	I	O	S	E
R	R	Q	P	O	S	T	M	U	U	P
A	C	U	H	Z	U	P	N	G	F	T
V	U	X	G	L	H	E	H	B	M	U
I	R	Z	O	T	V	X	M	E	X	N
T	Y	V	R	I	F	E	S	A	R	E
Y	E	A	U	R	A	N	U	S	R	E
R	E	U	J	U	P	I	T	E	R	S

CLUES

1. Gases that surround a planet _____
2. Third planet from the Sun. _____
3. A force that attracts matter toward the center of the Earth. _____
4. The Great Red Spot is found on this planet. _____
5. This Red Planet is the subject of many fictional stories. _____
6. Caloris Basin can be found on this planet. _____
7. The only satellite of the Earth. _____
8. This plantet was not discovered until 1846. _____
9. The smallest and coldest planet of all. _____
10. To move in an orbit around an object. _____
11. On complete turn of a planet on its axis. _____
12 The planet with the most satellites. _____
13. A bluish-green gas planet discovered in 1781. _____
14. The sister planet of the Earth. _____

© 1996 Kelley Wingate Publications   CD-3728

Name _____ Skill: Planets

TRUE OR FALSE?

Read each statement carefully. If the statement is true, put a T on the blank. If the statement is false, put a F on the blank.

_____ 1. Mars has three moons: Phobos, Deimos, and Olympus.

_____ 2. The Voyager 2 spacecraft passed over Neptune in 1986.

_____ 3. Mars is a rocky planet that is much warmer and larger than Earth.

_____ 4. Caloris Basin, a large crater area, is found on Mercury.

_____ 5. The Great Red Spot is a huge storm on Neptune.

_____ 6. Miranda, Ariel, Umbriel, Titania, and Oberon moons of Uranus.

_____ 7. Mercury is composed of gases and has no solid surface.

_____ 8. Venus appears bright because the clouds reflect so much light.

_____ 9. The soil on Mars contains potassium which makes it appear red.

_____ 10. The canyon called Valles Marineris is found on Mercury.

_____ 11. Neptune was discovered by two astronomers in 1846.

_____ 12. Uranus has more rings than any other planet.

_____ 13. Jupiter has more satellites than any other planet.

_____ 14. Mars is the only planet, other than Earth, that can support life.

_____ 15. The tilt of the Earth's axis allows this planet to have seasons.

_____ 16. Mercury is the second planet from the Sun.

_____ 17. Charon is a moon that is almost as large as the planet it orbits.

_____ 18. Pluto is covered with frozen gases that make an unusual snow.

© 1996 Kelley Wingate Publications

Name _____ Skill: Planets

Who Am I?

Read each clue and decide who or what is the answer. Write your answer on the line. Answers may be used more than once!

1. **Triton is one of my many moons.** _____

2. **I have 8 rings and at least 24 satellites.** _____

3. **I was named for the Greek god, Ares.** _____

4. **I have no rings and am the planet closest to the Sun.** _____

5. **I am 150 million km. away from the Sun.** _____

6. **The iron in my soil makes me look red.** _____

7. **I am the first of the four gas planets.** _____

8. **I am usually the ninth planet, but sometimes I am eighth.** _____

9. **I am the only celestial body (besides Earth) that man has walked on.** _____

10. **I rotate from West to East.** _____

11. **My surface is covered with a pudding-like snow.** _____

12. **I have 15 satellites, the largest of which is Miranda.** _____

13. **I am flattened at my poles because I rotate so quickly (10 hrs , 40 mins).** _____

14. **I have mountains, valleys, plains, and continents like the Earth, but I do not have water.** _____

© 1996 Kelley Wingate Publications   CD-3728

Name _____   Skill: Planets

Rank the Planets
List the planets in order for the information given.

Alphabetical Order	Size (Largest to Smallest)	Temperature (Hottest at any time)
1. _____	1. _____	1. _____
2. _____	2. _____	2. _____
3. _____	3. _____	3. _____
4. _____	4. _____	4. _____
5. _____	5. _____	5. _____
6. _____	6. _____	6. _____
7. _____	7. _____	7. _____
8. _____	8. _____	8. _____
9. _____	9. _____	9. _____

Rotation (Length of Day)	Revolution (Length of Year)	Distance From Sun
1. _____	1. _____	1. _____
2. _____	2. _____	2. _____
3. _____	3. _____	3. _____
4. _____	4. _____	4. _____
5. _____	5. _____	5. _____
6. _____	6. _____	6. _____
7. _____	7. _____	7. _____
8. _____	8. _____	8. _____
9. _____	9. _____	9. _____

© 1996 Kelley Wingate Publications   CD-3728

Name _____ Skill: Planets

LABEL THE MAP
Label the planets and Sun correctly.

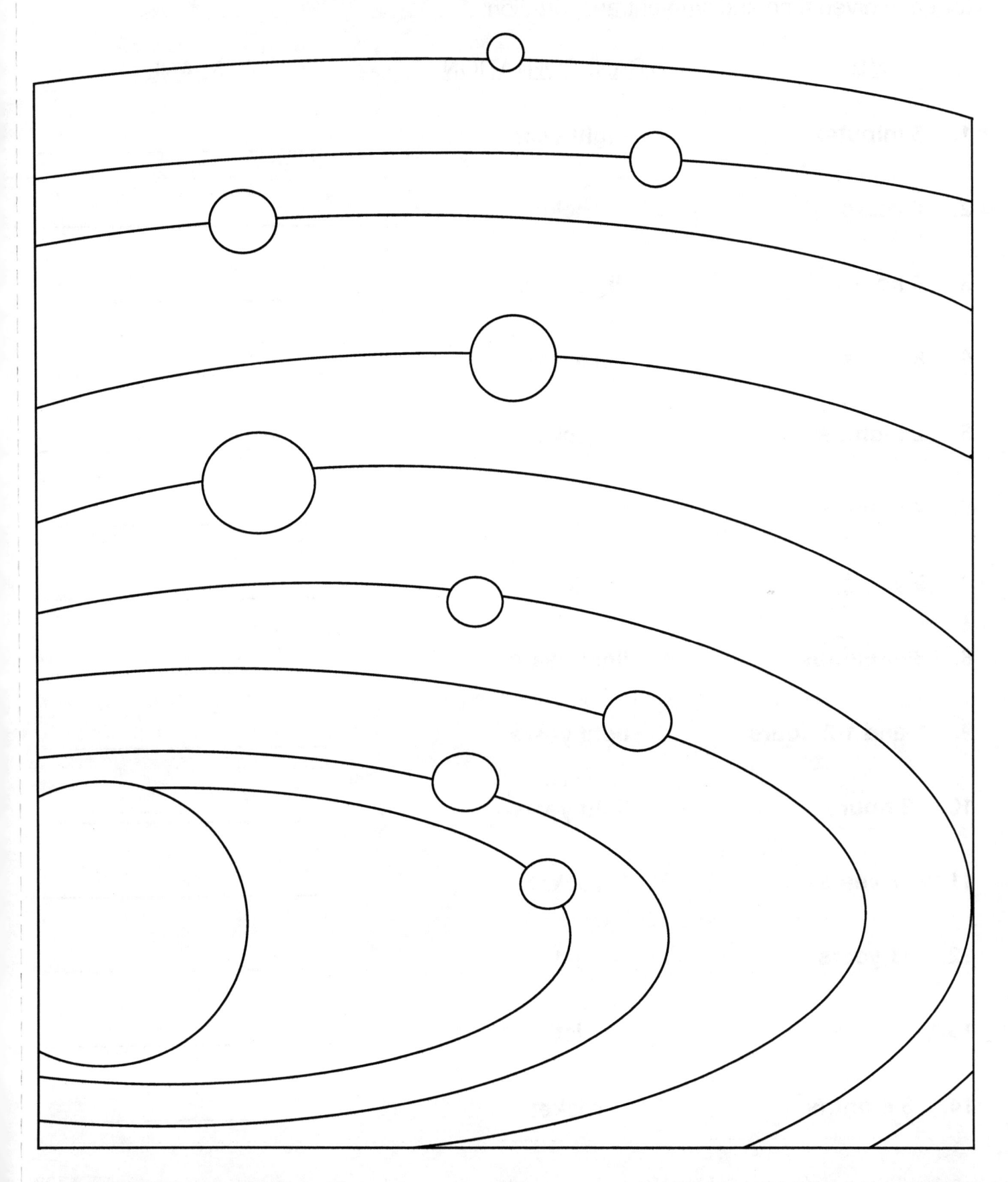

© 1996 Kelley Wingate Publications   37   CD-3728

Name _____   Skill: Planets

Plan A Vacation

You are planning a vacation on another planet. Name the furthest planet you can visit for each given time and type of transportation.

TIME	TRANSPORTATION	PLANET
1. 5 minutes	light years	_____
2. 4 years	rocket	_____
3. 6 hours	light years	_____
4. 8 years	rocket	_____
5. 2 months	rocket	_____
6. 4 months	jet	_____
7. 9 years	jet	_____
8. 35 minutes	light years	_____
9. 1 and 1/2 hours	light years	_____
10. 3 hours	light years	_____
11. 17 years	rocket	_____
12. 11 years	jet	_____
13. 6 years	jet	_____
14. 3 months	rocket	_____

© 1996 Kelley Wingate Publications   CD-3728

Name _____ Skill: Planets

Matching Planets

Match the planets to the facts about them.

PLANET   **FACT**

A. Mercury

_____ I am a mostly blue planet protected by a layer of clouds.

B. Venus

_____ I have 11 rings and 15 satellites.

C. Earth

_____ My eight rings reach into space about the same distance as it is between the Earth and the Moon.

D. Earth's Moon

E. Mars

_____ Triton and Nereid are two of my satellites.

F. Jupiter

_____ I am named for the Roman god of death.

G. Saturn

H. Uranus

_____ I have the Valles Marineris, a huge canyon, on my surface.

I. Neptune

J. Pluto

_____ Charon, my moon, is almost as large as I am.

_____ I am tilted on my axis, which gives me seasons.

_____ I am a gas planet with only three rings.

_____ My rotation is about 243 Earth days.

_____ My dark side is never facing the Earth.

_____ My revolution and rotation are both 27 days.

© 1996 Kelley Wingate Publications   CD-3728

Name _____   Skill: Stars

What is a star?

Somewhere in space a cloud of gas (mainly hydrogen and helium) and dust begins to collect. This cloud is called a nebula (a star in the making). For a few billion years the cloud contracts and grows warmer as more gas and dust are pulled in. When the protostar is dense enough the pressure and heat cause energy releasing reactions (a sort of fiery explosion of atoms) and the cloud begins to burn. The cloud is now a young star. The hydrogen fuses to form helium and energy is radiated as the star expands and burns brightly for another few billion years. As the hydrogen disappears, the star grows older it begins to cool and becomes a red giant (giving off lots of light but not much energy). Smaller stars become white dwarfs and larger stars explode as supernovae (burning brightly and quickly) and finally dies. If the helium core survives the explosion it may become a black hole.

1. What are the stages a star goes through during its lifetime?
    1. A cloud of gases and dust
    2. _____
    3. _____
    4. _____
    5. _____
    6. dies or becomes a black hole

2. What are the two main gases found in stars?

_____

3. What is a red giant?

_____

_____

4. What two things cause the nebula gases to begin burning?

_____

© 1996 Kelley Wingate Publications   CD-3728

Name _____ Skill: Stars

Star Color

If you have ever watched a fire, you have seen that parts of it are different colors. Scientists know from studying heat that different colors are produced by different degrees of temperature. Stars burning at different temperatures will also produce various colors. The coolest stars are red (also called M-type stars) with surface temperatures of about 3,038 °C (5,500° F). Our Sun is a G-Type star, light orange in color, with a temperature of 5,540° C (10,000°F). Yellow stars have temperatures of about 6,650°C 12,000°F) and are known as F-type. At about 6,900°C (12,500°F) stars turn white and are called A-type stars. The hottest stars are blue, burning between 9,400 and 20,500°C (17,000 and 37,000°F). These blue stars are called O-type.

1. Is it possible for any life to exist on a star? Why or why not?

_____

_____

2. What causes stars to have different colors?

_____

_____

3. Complete the chart below by filling in the correct information.

STAR CHART				
COOLEST ←	- -	- - - - - - - - - - - - - - - - - - -	→	HOTTEST
Color:	Color:	Color:	Color:	Color:
Temperature:	Temperature:	Temperature:	Temperature:	Temperature:
C:	C:	C:	C:	C:
F:	F:	F:	F:	F:

© 1996 Kelley Wingate Publications   41   CD-3728

Name _____ Skill: Stars

Star Size

Although most stars are about the same size as our Sun, stars do have different sizes. A few are dwarf stars which are much smaller than the Earth. Others are supergiant stars that are hundreds of times larger than our Sun. It is difficult for scientists to measure the diameter of stars, but they have developed several different ways to determine an approximate size. Our Sun is a medium-sized star with a diameter of about 1,400,000 km (865,000 miles). It would take over 100 Earth's to make the diameter of the Sun. Antares is a supergiant star with a diameter about 330 times larger than our sun! One of the smallest stars known is Van Maanen's Star which is about 9,800 km. (6,100 mi.) wide. That is about the size of the planet Mars.

1. What is a diameter? Use the dictionary and write the answer in your own words.

_____

_____

_____

2. What is the diameter of the Sun?

   _____ kilometers

   _____ miles

3. Which star is about 330 times larger than the Sun?

_____

4. Using the information from the paragraph, correctly label these three stars as Antares, Van Maanen, and Sun:

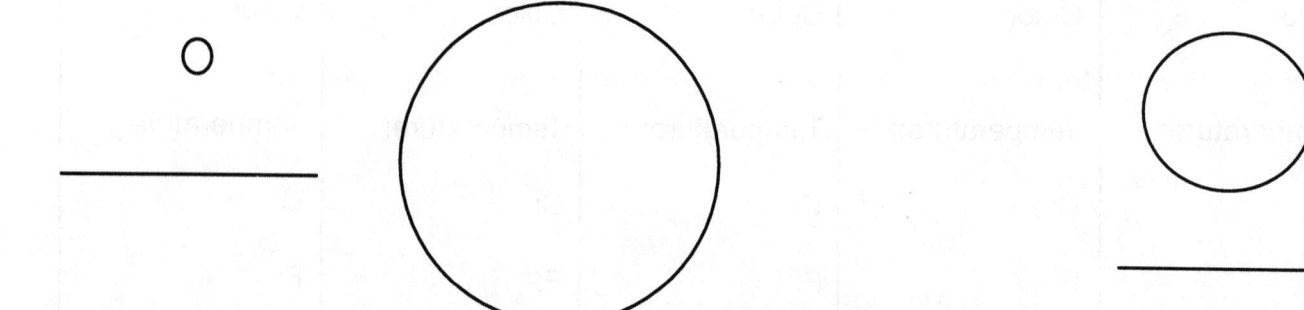

© 1996 Kelley Wingate Publications   42   CD-3728

Name _____ Skill: Stars

Life Span

Human beings have a life span of about 80 years, but a star usually lives for about ten billion years! The first stage is called a nebula - a cloud of hydrogen and dust. When the hydrogen compacts, changes, and explodes the nebula becomes a baby star. The third stage is a star. During this time the hydrogen continues to fuse into helium and radiates light and heat. At 4.6 billion years old, our Sun is considered a middle-aged star! As the star grows older the hydrogen begins to disappear and the helium grows. At the fourth stage, the helium core cools and shrinks rapidly as the outer shell glows brighter than before but giving off little radiant energy. Now the star becomes a giant or supergiant. During the fifth stage the supergiant cools and contracts until it becomes a white dwarf. This star is slowly dying now that the hydrogen and helium have been nearly used up. Finally, the white dwarf becomes a black dwarf, completely collapsing in on itself.

1. List the six stages in the life of a star:

1. _____
2. _____
3. _____
4. _____
5. _____
6. _____

2. What is the life span of most stars?

_____ years

3. How old is the Sun and what stage is it in?

_____ years

_____ stage

© 1996 Kelley Wingate Publications   43   CD-3728

Name _____ Skill: Planets

TRUE OR FALSE?

Read each statement carefully. If the statement is true, put a T on the blank. If the statement is false, put a F on the blank.

_____ 1. A star is a burning ball of gases and dust.

_____ 2. The hottest stars are deep red in color.

_____ 3. A baby star is a cloud of hydrogen and dust.

_____ 4. Antares is one of the largest known stars.

_____ 5. A white dwarf is a star very near the end of its life.

_____ 6. The life span of a star is about three million years.

_____ 7. Our Sun is a G-type star.

_____ 8. The blue stars have a temperature up to 20,600° C (37,000° F).

_____ 9. A nebula is a star that has burned out.

_____ 10. The van Maanen star is a dwarf.

_____ 11. Most of the stars are supergiants.

_____ 12. As hydrogen changes into helium it explodes as fiery energy.

_____ 13. The core of a giant or supergiant star is mainly helium.

_____ 14. The last stage of a star's life is black dwarf.

_____ 15. At the giant stage, a star produces more energy and less light.

_____ 16. The coolest stars are red in color.

_____ 17. When a star dies, the helium core may become a black hole.

_____ 18. Our Sun is about 4.6 billion years old.

© 1996 Kelley Wingate Publications   CD-3728

Name _____   Skill: Sun

Our Sun

    The star nearest to the Earth and the center of our solar system is, of course, the Sun. Without its heat and light we could not survive. The Sun rotates on its axis (just like the planets) and averages 25 days for each rotation. The Sun is an average-sized star that is about half way through its ten billion year life. It is yellow in color and is classified as a G-type star. It is composed, or made up of, about 2/3 hydrogen and 1/3 helium. Energy is produced in the core, or middle, of the sun as hydrogen fuses (or alters) because of heat (2 million degrees C or 1 million degrees F) and pressure. The energy rises to the surface of the sun where it cools to about 18,100° C (10,000° F). From the surface the energy is radiated into space as heat and light.

1. Why is the Sun so important to life on Earth?

_____

_____

2. How long is one solar day (one rotation of the Sun)?

_____

3. Where is the Sun's energy produced?

_____

4. What is the temperature of the surface of the Sun?

_____ **Celsius**   _____ **Fahrenheit**

5. What would happen to life on Earth if we were any closer or further away from the sun?

_____

_____

_____

_____

© 1996 Kelley Wingate Publications   CD-3728

Name _____   Skill: Sun

Solar Sunspots

As early as 1612 Galileo noticed several black spots on the Sun. These dark patches, called sunspots, are set back into the surface of the sun and are actually cooler than the surrounding area. Although dark in color, sunspots produce a lot of light. In fact, just one sunspot could light up a dark sky! They may look small compared to the size of the Sun, but sunspots often have a diameter of 18,000 km (11,000 mi.). Sunspots are usually found in groups and it is believed that they have a magnetic attraction toward each other. Scientists have also discovered that there is a cycle to sunspots. They reappear in the same places about every eleven years. Most sunspots form and disappear within a two week period, but some last up to a year.

1. What is a sunspot?

_____

_____

2. How often does the sunspot cycle repeat itself?

_____

3. Why do sunspots usually form in groups?

_____

_____

4. What is the average diameter of a sunspot?

_____ km   _____ mi

5. How long does it normally take for a sunspot to form and disappear?

_____

6. What is the longest period of time a sunspot might last?

_____

© 1996 Kelley Wingate Publications

Name _____ Skill: Sun

Solar Flares

Solar flares are sudden burst of energy in the form of fire that erupts on the Sun's surface. These powerful fire bursts usually occur over an area of a sunspot. No one knows for certain what causes solar flares, but it is believed they are a form of magnetic storm created by solar winds and sunspots. These fire storms can be quite large, covering up to one billion square kilometers (386 million square miles). Usually a flare happens in less than a minute, but it can give off an extreme amount of radiation (waves of energy). This radiation can be harmful to astronauts in space and can even disturb radio waves or compasses on Earth! Scientists track solar flares on the Sun, but they have seen flares on other stars as well.

1. Which are more dangerous to man, sunspots or solar flares? Why?

_____

_____

_____

2. What causes solar flares?

_____

_____

3. How are sunspots and solar flares related?

_____

_____

_____

4. How large can a solar flare become?

_____ square km

_____ square mi

© 1996 Kelley Wingate Publications   CD-3728

Name _____   Skill: Sun

Solar Prominences

The most spectacular event on the Sun is undoubtedly a solar prominence. These fantastic happenings are great rose colored gas clouds that quickly shoot up from the Sun's surface for thousands of kilometers, then loop over and return to the surface. A typical prominence will rise about 40,000 km (25,000 mi) above the surface of the Sun. Solar prominences usually take about an hour from start to finish, although some very large ones have been known to last for several months. Scientists do not fully understand what causes these happenings, but they believe the Sun's magnetic field has something to do with them.

1. The word "solar" means sun and the word "prominent" means immediately noticeable. Why do you think scientists named these shooting fire clouds solar prominences?

_____

_____

_____

2. In your own words, what is a solar prominence?

_____

_____

_____

3. Why do you think prominences are probably more exciting to see than sunspots or solar flares?

_____

_____

_____

4. How long does a typical solar prominence last?

_____

© 1996 Kelley Wingate Publications   CD-3728

Name _____ Skill: Sun

Solar Eclipse

The Sun is about 400 times larger than the Earth's Moon. It is also about 400 times further from the Earth than the Moon. This strange coincidence makes the Sun and the Moon appear to be about the same size when viewed from the Earth. Every so often the Sun and the Moon are in a direct line with some location on the Earth. Because they appear to be the same size, the Moon actually blocks out the Sun's light for a short time. The Earth becomes almost as dark as night as the Moon's shadow (about 160 km. or 100 mi. across) passes over the land. This event is called a total solar eclipse (all of the Sun is blocked). Other parts of the Earth may see the Moon block only part of the Sun as the two orbits cross paths. This is known as a partial solar eclipse (part of the Sun is blocked). If you were to live in the same city forever, you would see a total eclipse only once every 400 years!

1. What makes the Moon and the Sun appear to be the same size?

_____

_____

2. How wide is the shadow cast by the Moon during a total eclipse?

_____ kilometer

_____ miles

3. What is the difference between a total eclipse and a partial eclipse?

Total Eclipse _____

_____

Partial Eclipse _____

_____

4. How often does a total solar eclipse often in the same location?

_____

© 1996 Kelley Wingate Publications

Name _____ Skill: Stars

TRUE OR FALSE?

Read each statement carefully. If the statement is true, put a T on the blank. If the statement is false, put a F on the blank.

_____ 1. Our Sun is the largest star in our solar system.

_____ 2. Sunspots are bits of sun that fly off into space.

_____ 3. Stars are made of gases and bits of dust.

_____ 4. Sunspots have a higher temperature than the area around them.

_____ 5. Galileo was the first astronomer to see solar prominences.

_____ 6. Solar flares can affect radio signals and compasses on the Earth.

_____ 7. The shadow of the Moon during an eclipse is 16 km (10 mi.) wide.

_____ 8. The Moon is blocked by the Sun during a solar eclipse.

_____ 9. Solar prominences are loops of fire shot from the Sun's surface.

_____ 10. Flares can be seen on other planets.

_____ 11. Sunspots, although dark, produce a great amount of light.

_____ 12. Sunspots follow a cycle of 11 years.

_____ 13. Our Sun is very old and has become a supergiant.

_____ 14. The Sun is one of the smallest stars.

_____ 15. The Sun's magnetic field is probably the cause of flares.

_____ 16. Sunspots last for several hundred years.

_____ 17. Solar prominences spurt about 40,000 km. (25,000 mi.) high.

_____ 18. The Sun is about 400 times larger than the Earth's Moon.

© 1996 Kelley Wingate Publications   CD-3728

Name _____ Skill: Stars

ACROSTIC

Read the clues and figure out what they are naming. Write the word in the spaces provided. (Clue number one goes first and begins with an A.)

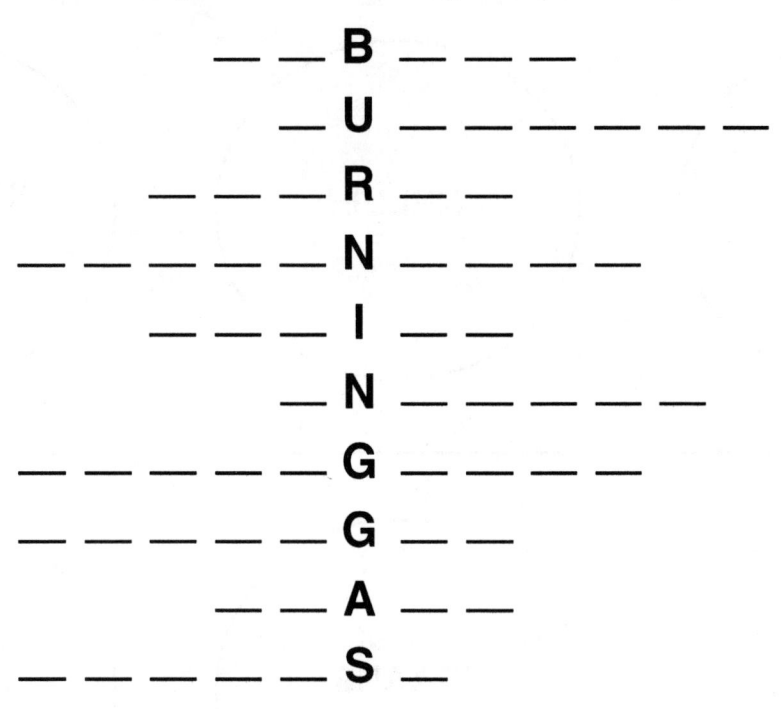

```
      _ _ B _ _ _
        _ U _ _ _ _ _ _
    _ _ _ R _ _
_ _ _ _ _ N _ _ _ _
      _ _ _ I _ _
        _ N _ _ _ _ _
  _ _ _ _ _ G _ _ _ _
  _ _ _ _ _ G _ _
        _ _ A _ _
  _ _ _ _ _ S _
```

1. A cloud of hydrogen, helium, and dust.

2. A group of dark spots on the surface of the Sun.

3. Small bursts of fire near sunspots.

4. Tall loop of fire bursting from the surface of the Sun.

5. A gas at the core of an older star.

6. One of the largest known stars.

7. A star near the end of its life that gives off lots of light but not much heat.

8. The gas that composes most of a baby star.

9. A star near the end of its life that has burned up most of its hydrogen and helium.

10. This happens when the Moon comes directly between the Sun and the Earth, casting a huge shadow on the Earth as it blocking out the Sun.

© 1996 Kelley Wingate Publications   CD-3728

Name _____   Skill: Stars

SOLAR SCRAMBLE

The Sun was so hot yesterday that it scrambled all the letters in the words below. Unscramble the letters and write the "star" words below each sun.

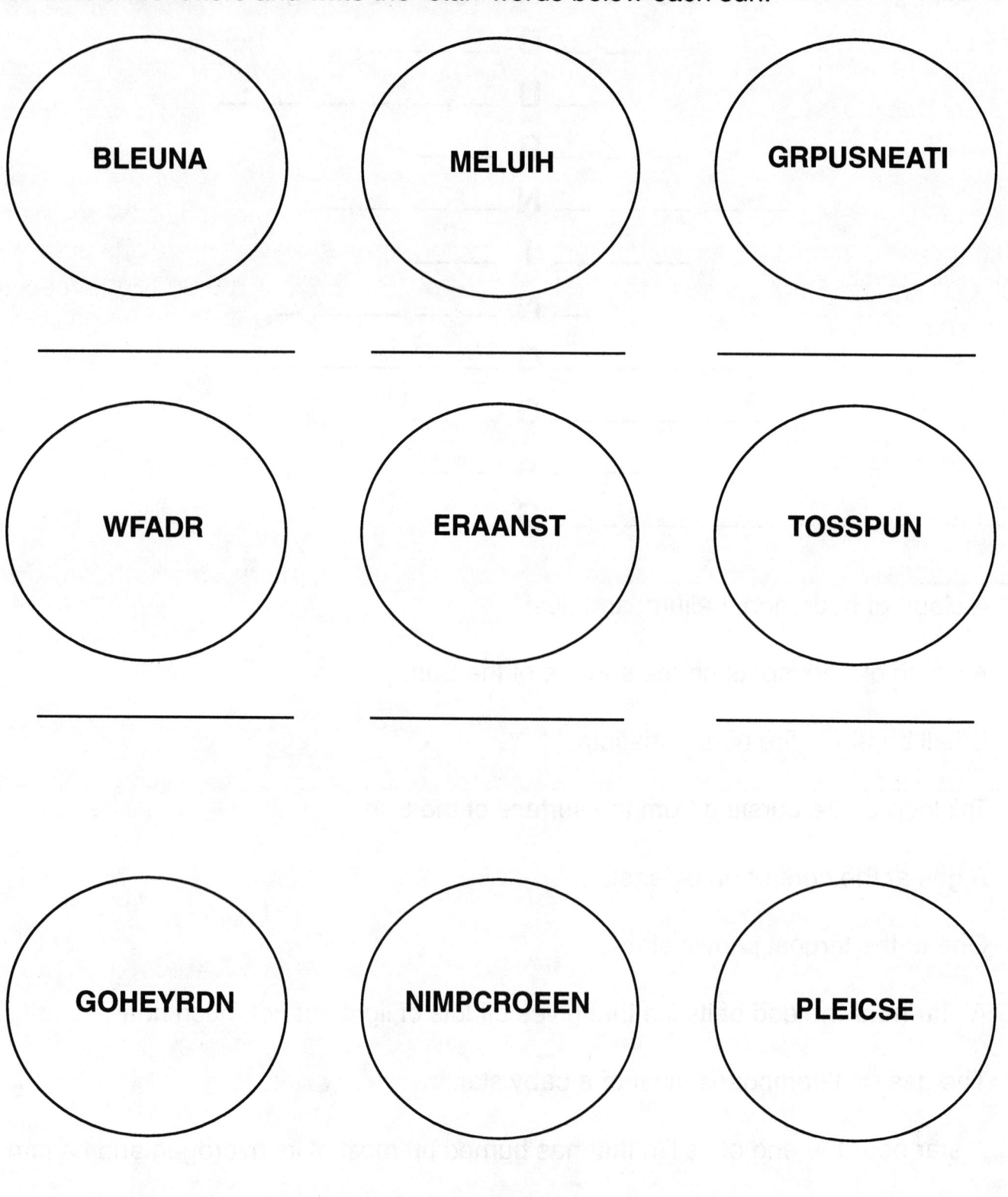

BLEUNA

MELUIH

GRPUSNEATI

_____   _____   _____

WFADR

ERAANST

TOSSPUN

_____   _____   _____

GOHEYRDN

NIMPCROEEN

PLEICSE

_____   _____   _____

© 1996 Kelley Wingate Publications   52   CD-3728

Name _____   Skill: Stars

The Milky Way

Our solar system (the Sun and planets) is very large to us, but it is only one small part of our universe. There are several billion other stars in space that we know very little about! When early astronomers looked at the sky they noticed a huge band of stars that looked somewhat like a road of milk. They began to call this band "gala" which is the Greek word for milk. From this term we developed the word "galaxy," which means a collection of stars. The band of stars has become known as the Milky Way, another name for our galaxy. The Milky Way is so large that it must be measured in light years (the distance a ray of light can travel in one year). Light travels at 300,000 km. (186,000 mi.) per second! The Milky Way is about 100,000 light years in diameter. If we fit a model of our galaxy on this paper, the Sun would be smaller than the dot at the end of this sentence and the Earth could only be seen through a microscope.

1. What is a galaxy?

_____

_____

_____

2. Why was our galaxy named after milk?

_____

_____

_____

3. What is a light year?

_____

_____

4. Why is the galaxy measured in light years rather than kilometers or miles?

_____

_____

_____

© 1996 Kelley Wingate Publications   CD-3728

Name _____   Skill: Stars

Novas and Supernovas

The Latin word "novus" means new. In the early days of astronomy a star which suddenly seemed to appear was called a nova (new star). Today we know that a nova is actually an old star that is burning out. The surface of the star suddenly explodes, creating a light that may be thousands of times brighter than the original star. After a few days, months, or years the star returns to its original brightness. This may happen several times before the star becomes a dwarf. There have been hundreds of novas in our galaxy during the past 100 years. Sometimes a star explodes completely, creating a glow that is billions of times brighter than the original star. These explosions are called supernovas, occurring only once every few hundred years. In 1987 a supernova was observed in another galaxy called the Large Magellanic Cloud.

1. Why did early astronomers call a suddenly glowing star a nova?

_____

_____

_____

2. What happens to a nova after it becomes so bright?

_____

_____

_____

3. List two ways in which novas and supernovas are different:

   1. _____

   2. _____

4. What happens to a supernova after it becomes so bright?

_____

_____

_____

© 1996 Kelley Wingate Publications　　54　　CD-3728

Name _____   Skill: Stars

Black Holes

    Imagine a star many times larger than our Sun. Its gravity is powerful, pulling in any mass (like gases and dust) that comes too near. The gravity of this star continues to grow until it becomes even stronger than the speed of light. At that point the star would collapse into itself, leaving a large black hole that would capture anything near it. As early as 1798 the theory was developed that these black holes could occur, but it has never been proven. The Hubble Space Telescope photographed a disk of dust and gas in the center of galaxy M87. This disk is believed to be a black hole, but no one knows for sure. Black holes are one of the many mysteries of space yet to be solved.

1. Why do scientists consider black holes to be mysteries?

_____

_____

2. Is a black hole really a hole? Explain.

_____

_____

_____

3. In which galaxy have scientist found what they think is a black hole?

_____

4. Explain this statement: Black holes are only theories (ideas).

_____

_____

_____

_____

© 1996 Kelley Wingate Publications   55   CD-3728

Name _____ Skill: Stars

NUMBER MATCHING

Match the numbers on the left with the facts about them on the right.

A. 186,000 miles per second   _____ temperature of the coolest star

B. 10 billion years   _____ diameter of the Milky Way Galaxy

C. 4.6 billion years   _____ life span of a star

D. 3,038° C (5,500° F)   _____ Galileo discovered sunspots

E. 20,500° C (37,000° F)   _____ height of a solar prominence

F. 1987   _____ rotation and revolution of the Moon

G. 100,000 light years   _____ size of Antares

H. 2,000,000° C (1,100,000° F)   _____ supernova discovered in another galaxy

I. 1798   _____ temperature at the core of the Sun

J. 330 times larger than the Sun   _____ theory of black holes was developed

K. 11 years   _____ size of the Moon

L. 2/3 hydrogen, 1/3 helium   _____ speed of light

M. 1/4 the size of the Earth   _____ cycle for sunspots

N. 1612   _____ amount of gases in the Sun

O. 27 days   _____ temperature of the hottest stars

P. 160 km (100 mi)   _____ age of our Sun

Q. 40,000 km (25,000 miles) above the Sun   _____ diameter of Moon's shadow during total eclipse

© 1996 Kelley Wingate Publications 56 CD-3728

Name _____ Skill: Stars

TRUE OR FALSE?

Read each statement carefully. If the statement is true, put a T on the blank. If the statement is false, put a F on the blank.

_____ 1. The Milky Way Galaxy is 100,000 light years in diameter.

_____ 2. The Sun and its planets are one galaxy.

_____ 3. Each galaxy has millions of stars.

_____ 4. "Gala" is the Greek word for milk.

_____ 5. Our solar system is part of the Milky Way Galaxy.

_____ 6. "Novus" or nova means old.

_____ 7. A nova is a star that is just beginning its life.

_____ 8. Scientists have seen hundreds of novas in the last 100 years.

_____ 9. A nova may shine brightly then fade several times before it dies.

_____ 10. Supernovas are stars that explode completely.

_____ 11. In 1987, scientists found a supernova in galaxy M87.

_____ 12. There are at least three black holes that scientists have studied.

_____ 13. Black holes are stars that absorb their own light.

_____ 14. Scientists discovered a black hole in 1798.

_____ 15. The Hubble Telescope photographed what could be a black hole.

_____ 16. Black holes are only unproven theories.

_____ 17. The gravity of a black hole is not as strong as the speed of light.

_____ 18. A black hole is a large hole.

© 1996 Kelley Wingate Publications   CD-3728

Name _____ Skill: Stars

STAR CROSSWORD

ACROSS

2. A gas that makes up the majority of a new star.
3. These stars are red and the coolest.
4. When the Moon has moved directly between the Sun and the Earth.
7. An exploding star is called this.
8. The name of the largest known star.
11. What we call a vast group of stars.
13. A star that has collapsed on itself.
14. The name of our galaxy.

DOWN

1. Loops of fire that burst from the Sun.
2. A gas that makes up the majority of an older star.
5. The fourth stage of a star (when it glows brightly).
6. Bursts of solar energy that occur over or near a sunspot.
9. A cloud of hydrogen, dust, and helium.
10. The name of the star in our solar system.
12. An old star that flares brightly as it begins to die.

© 1996 Kelley Wingate Publications   58   CD-3728

Name _____ Skill: Theory

STARS

Read the clues below and give the name of what it describes.
Circle the names in the word search. They may go in any direction.

S	F	H	Y	D	R	O	G	E	N	F
O	N	R	Y	X	A	L	A	G	L	P
E	T	O	A	N	W	S	K	A	R	S
C	A	Y	I	W	D	M	R	O	H	U
L	L	D	P	L	D	E	M	X	P	N
I	U	J	V	E	L	I	A	F	X	S
P	B	J	I	R	N	S	I	P	P	
S	E	Y	A	E	U	S	B	U	O	O
E	N	V	N	I	Q	Y	O	N	N	T
Y	O	C	U	M	M	U	I	L	E	H
N	E	X	S	E	R	A	T	N	A	T

CLUES

1. The largest known star. _____
2. An older star that has become small. _____
3. This happens when the Sun, Moon, and Earth are in a direct line. _____
4. This gives off radiation that disturbs radio waves on Earth. _____
5. What the Milky Way is. _____
6. The main gas in older stars. _____
7. The main gas in new stars. _____
8. The 1st stage of a star. _____
9. An old star that suddenly explodes. _____
10. Hot blue stars. _____
11. The most spectacular event on the Sun. _____
12. A middle aged G-type star in our solar system. _____
13. These reappear in the same place on the Sun every eleven years. _____
14. The average number of years a star exists. _____

© 1996 Kelley Wingate Publications

Name _____ Skill: Celestial Sights

Constellation

Early Greeks noticed that the stars seem to form patterns that make it easy to describe directions or areas of the sky. With a little imagination, groups of stars became pictures. We call these "pictures" constellations. Ptolemy named 48 different groups of stars (mainly characters from Greek mythology). In time, other astronomers added forty more so now every star is part of a constellation. The sky is divided into sections called northern, equatorial (over the equator), and southern. The constellations give astronomers a kind of map of the stars, making it easier to locate any one of of them!

1. What is a constellation?

_____

_____

2. How do constellations make it easier to locate stars?

_____

_____

3. Name the three sections the sky is divided into:

   1. _____

   2. _____

   3. _____

4. How many constellations are there?

_____ constellations

© 1996 Kelley Wingate Publications   CD-3728

Name _____ Skill: Celestial Sights

URSA MAJOR

One constellation that is easy to locate is Ursa Major, the Great Bear. It is commonly called the Big Dipper and can be seen in the northern skies all year round. Ursa Major is a group of seven very bright stars that form an almost rectangular shape with a bent handle of three stars. The first two stars in the bowl, Dubhe and Merak, are called Pointers. If you draw an imaginary line from the lower star (Merak) through the upper star and continue until you come to a very bright star, you have located Polaris (the Pole Star or North Star). Polaris remains at the same place in the sky all the time and is very bright. Ancient sailors, among others, used this star as a guide so they always knew the direction in which they were headed.

1. What are two other names for Ursa Major?

    1. _____

    2. _____

2. Why are Dubhe and Merak called Pointers?

_____

_____

_____

3. What are two other names for Polaris?

    1. _____

    2. _____

4. Why is Polaris an important star?

_____

_____

© 1996 Kelley Wingate Publications   CD-3728

Name _____ Skill: Celestial Sights

Comets

A comet is a celestial body that orbits the Sun, passing quickly amongst planets on its journey. It appear as a flat ball, sometimes dragging a tail behind it. A comet is made of several gases, water, and dust that are frozen into a kind of dirty snowball. The average diameter of a comet is 100,000 km. (62,000 mi.). When the comet is far from the Sun it travels at about 3,300 km. (2,000 mi.) per hour. As it nears the Sun the speed may increase to over 161,000 km. (100,000 mi.) per hour. About ten new comets are discovered every year. You can usually see a comet without using a telescope about once every three years.

1. What is a comet made of?

_____

_____

2. Compare a planet and a comet. How are they alike? Different?

   Alike _____

_____

   Different _____

_____

3. Does a comet always travel at the same speed? Explain.

_____

_____

_____

4. Pretend that you are the first person to discover a new comet. What would you name it and why?

_____

_____

© 1996 Kelley Wingate Publications   CD-3728

Name _____ Skill: Celestial Sights

Parts of a Comet

A comet is much like a dirty ice ball. The center, or nucleus, of this ball is composed of frozen gases, water, and dirt. As it travels through space, the outside of the nucleus warms and gives off a cloud of gas that surrounds it. This cloud is called the comet and averages 100,000 km. (62,000 mi.) in diameter. As the comet nears the Sun it travels even faster and the coma becomes much larger. The solar winds push against the coma, creating a tail that streams from the nucleus. As the comet nears the Sun, the tail is behind it. However, as the comet is moving away from the Sun, the tail is in front! Scientists believe the strong solar winds are responsible for this fact.

1. Name the three parts of a comet.

    1. _____

    2. _____

    3. _____

2. Label the three parts of the comet:

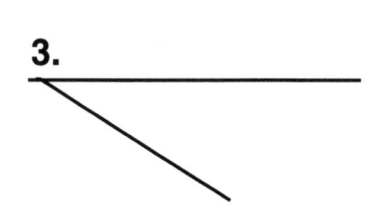

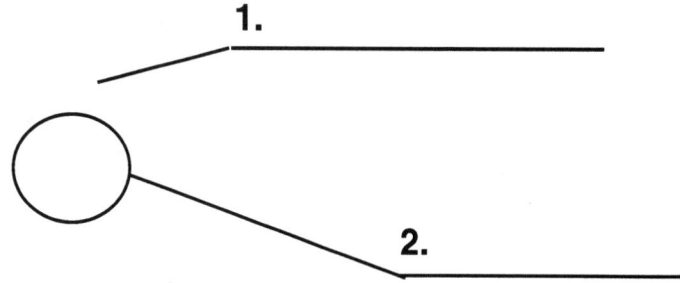

    1. _____

    3. _____

    2. _____

3. Why does the tail always point away from the Sun?

_____

_____

4. How long is a comet's tail?

_____

© 1996 Kelley Wingate Publications

Name _____ Skill: Celestial Sights

Halley's Comet

In 1682 Sir Edmund Halley saw a comet. He developed the theory that comets orbit the Sun, as planets do, and that this comet was the same one sighted in 1531 and 1607. Halley predicted that the comet would return in 1758. It did, and the comet was named Halley's Comet in honor of Sir Edmund. Halley's comet appears in our sky every 76 years. The comet makes an orbit of 7 billion miles that takes it just past Neptune. The tail of Halley's Comet has been measured at about 150 million km. (93 million mi.) which is the same distance as the Earth to the Sun!

1. Why is this comet called Halley's Comet?

_____

_____

2. How often is this comet seen by Earth?

_____

_____

3. How many times has Halley's Comet been visible to Earth since the year 1758? (Hint: How often does it appear?)

_____ times

4. How long is the tail of this comet?

_____ kilometers   _____ miles

5. In what year will Halley's Comet return to the Earth's sky?

_____

© 1996 Kelley Wingate Publications   CD-3728

Name _____ Skill: Celestial Sights

Asteroids

Asteroids, sometimes called minor planets, are small solid objects that also orbit the Sun. They are made of iron, nickel, stone, or any combination of these. There are thousands of asteroids and over 5,000 of them are found between Mars and Jupiter in what is called the Asteroid Belt. Asteroids range in size from many kilometers to a few meters. The largest known asteroid is called Ceres and measures 785 km. (500 mi.) in diameter. It is believed that asteroids are actually parts of the solar system that never joined together as planets. Most of the ones known today are probably pieces of larger ones that smashed against each other and broke apart. Scientists have found evidence of the Earth being hit by asteroids about the time of the ice age (2.3 million years ago). In more recent years asteroids have passed as close as 770,000 km. (478,000 mi.), but there have been no collisions.

1. Why are asteroids called minor planets?

_____

_____

_____

2. Where can we find the Asteroid Belt?

_____

_____

3. What are asteroids?

_____

_____

4. Look in the encyclopedia and find the names of two other large asteroids.

1. _____

2. _____

© 1996 Kelley Wingate Publications   CD-3728

Name _____   Skill: Celestial Sights

Meteoroids, Meteors, and Meteorites

Meteoroids are small bits of rock and dust that are found in space. They are often pulled into the Earth's atmosphere by the force of gravity. Once a meteoroid enters the atmosphere it begins to burn, creating a streak of light in the sky. At this point it is called a meteor, although it is commonly called a falling or shooting star. Although many meteors burn completely before they reach the ground, not all of them do. If a meteor reaches the ground it is called a meteorite. About 500 large meteorites make it to the ground each year, but only about 5 of those are ever recovered.

1. Describe the differences among these three words:

    meteoroid _____

    _____

    meteor _____

    _____

    meteorite _____

    _____

2. Why are meteors sometimes called falling stars?

    _____

    _____

3. What causes a meteoroid to enter the Earth's atmosphere?

    _____

4. Meteorites have caused craters on the Moon. Why do you think the Earth does not have as many as the Moon?

    _____

    _____

© 1996 Kelley Wingate Publications

Name _____   Skill: Celestial Sights

Meteor Showers

Almost any night, especially after midnight, it is possible to see a couple meteors as they fall to Earth. Sometimes so many fall at once that it is called a meteor shower. This happens when the Earth passes through a wave of meteoroids, believed to be left by a passing comet. The average meteor in such a shower is about the size of a grain of sand. Meteor showers are named after the constellation from which they seem to fall. The most brilliant shower is named Leonid because it seems to come from the constellation Leo, the lion. In 1966 over 1000 meteors per second were seen in the Leonid meteor shower!

1. What causes a meteor shower?

_____

_____

_____

2. How are meteor showers named?

_____

_____

_____

3. What is the average size of a meteor in a shower?

_____

4. What was so special about the Leonid Meteor Shower of 1966?

_____

_____

5. Where do the bands of meteoroids come from?

_____

_____

© 1996 Kelley Wingate Publications   CD-3728

Name _____ Skill: Celestial Sights

TRUE OR FALSE?

Read each statement carefully. If the statement is true, put a T on the blank. If the statement is false, put a F on the blank.

_____ 1. Ptolemy named all of the 88 constellations.

_____ 2. Ursa Major is the largest asteroid ever discovered.

_____ 3. Polaris is part of the Big Dipper constellation .

_____ 4. The North Star does not change position in our sky.

_____ 5. Dubhe and Merak are constellations.

_____ 6. A comet increases in speed as it nears the Sun.

_____ 7. A comet tail always points toward the Sun.

_____ 8. Comets are made of gases, water, and dust.

_____ 9. About 100 new comets are discovered each year.

_____ 10. Nucleus, coma, and tail are parts of a comet.

_____ 11. The heat from the Sun pushes the comet tail backward.

_____ 12. Halley's comet appears every 96 years.

_____ 13. Asteroids are made of nickel, iron, or stone.

_____ 14. The Asteroid Belt is found between Jupiter and Saturn.

_____ 15. Meteoroids are meteors that hit the ground.

_____ 16. Meteoroids are probably left in space by passing comets.

_____ 17. Meteroids are also called shooting stars.

_____ 18. Less than five meteorites make it to the ground each year.

© 1996 Kelley Wingate Publications

Name _____ Skill: Celestial Sights

Who Am I?

Read each clue and decide who or what is the answer. Write your answer on the line. Answers may be used more than once!

1. I named 48 of the constellations. _____

2. I can be seen every 76 years. _____

3. I am sometimes called a minor planet. _____

4. I am a small particle that burns as I fall to Earth. _____

5. I am a group of stars that form a picture. _____

6. I am made of frozen ice, gas, and dust. _____

7. I am the largest known asteroid. _____

8. I am small pieces of rock and dust left in space by a comet. _____

9. In 1682 I saw a fuzzy star with a tail and predicted when it would be seen again. _____

10. I am the core of ice and dust in a comet. _____

11. I was once a meteoroid and will soon be a meteorite. _____

12. I am what causes a comet's tail to point away from the Sun. _____

13. The speed of my orbit increases as I come close to the Sun. _____

14. There are 88 of us and most of us were named for characters in Greek mythology. _____

© 1996 Kelley Wingate Publications   CD-3728

Name _____ Skill: Celestial Sights

ACROSTIC

Read the clues and figure out the name of the astronomer. Write his name in the spaces provided. (Clue number one goes first, two second, and so on.)

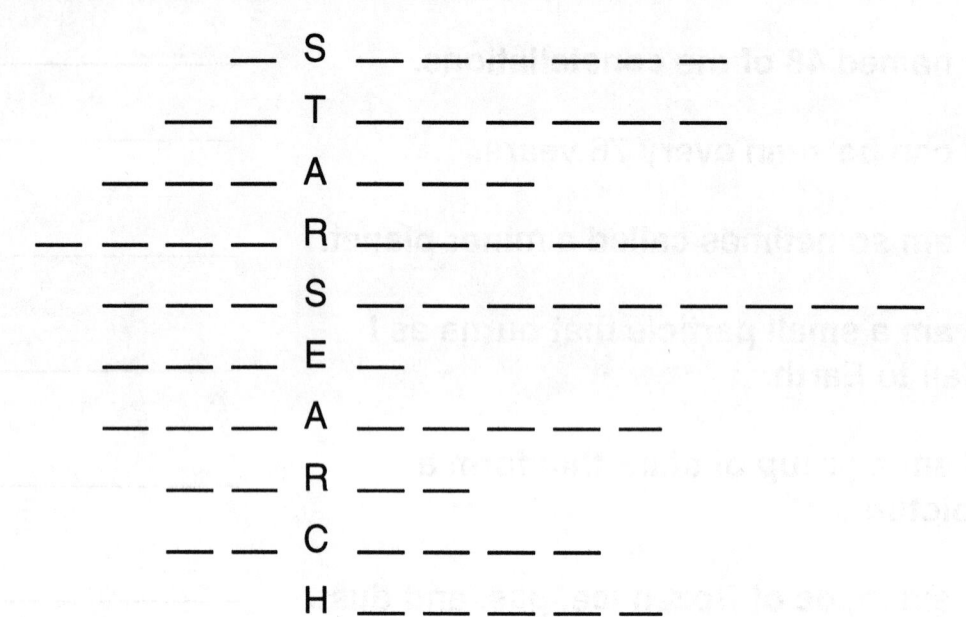

```
    _ S _ _ _ _ _ _
  _ _ T _ _ _ _ _
_ _ _ A _ _ _
_ _ _ _ R _ _ _ _ _ _
_ _ _ S _ _ _ _ _ _ _ _ _
    _ _ _ E _
  _ _ _ A _ _ _ _
    _ _ R _ _
    _ _ C _ _ _ _
      H _ _ _ _
```

1. Iron, nickel, or stone pieces that orbit the Sun.

2. A small bit of burned rock or dust that falls to the surface of the Earth.

3. The North Star used by early sailors to figure out directions at night.

4. A force that pushes the tail of a comet away from the Sun.

5. A group of stars that form a picture.

6. A ball of ice that orbits the Sun.

7. Another name for the Big Dipper or Great Bear.

8. The name of the largest known asteroid.

9. The frozen center of a comet that is surrounded by the coma.

10. The man who predicted a 76 year cycle for a particular comet.

© 1996 Kelley Wingate Publications   CD-3728

Name _____ Skill: Celestial Sights

CELESTIAL SIGHTS

Read the clues below and give the name of what it describes.
Circle the names in the word search. They may go in any direction.

N	T	C	N	A	V	M	T	Z	O	S	Z	C	U	Z
N	F	A	Z	I	T	X	Q	S	E	C	P	U	R	Y
Z	S	C	I	I	F	Y	D	R	W	R	O	S	S	O
Z	A	T	O	L	H	I	E	Z	R	E	L	J	A	L
Z	V	B	T	N	O	C	A	Y	W	I	R	C	M	Q
C	Z	N	Q	R	S	C	I	C	O	U	M	P	A	M
F	W	G	E	A	C	T	O	G	E	W	E	T	J	E
F	F	T	M	S	E	Y	E	M	A	F	T	O	O	T
V	S	O	E	I	P	C	F	L	E	X	E	L	R	E
A	C	V	K	L	D	U	N	E	L	T	O	E	E	O
G	U	D	N	S	U	M	N	F	N	A	R	M	S	R
O	N	K	Q	T	H	A	L	L	E	Y	T	Y	Y	O
P	V	G	X	H	P	O	L	A	R	I	S	I	A	I
G	A	O	D	S	O	L	A	R	W	I	N	D	O	D
M	E	T	E	O	R	I	T	E	Q	S	K	Y	X	N

CLUES

1. Solid objects that orbit the Sun in a beltway between Jupiter and Mars. _____
2. The largest known asteroid. _____
3. The cloud that surrounds a comet. _____
4. A ball of frozen gas, water, and dust. _____
5. A group of stars that form a picture. _____
6. He predicted when a comet would return. _____
7. "Falling stars". _____
8. A meteor that hits the surface of the Earth. _____
9. Small bits of rock or dust found in space. _____
10. The North Star. _____
11. The man who named many of the constellations. _____
12. The force that pushes a comet tail away from the Sun. _____
13. The part of a comet that streams behind the head. _____
14. A constellation that looks like a water dipper. _____

© 1996 Kelley Wingate Publications

Name _____ Skill: Space Exploration

History of Rockets

    Throughout time man has dreamed of space travel. Many cultures have folk tales and stories that tell of gods coming from the stars. Science fiction books are full of tales about travel to other planets and star systems. The rocket seemed the most likely way to turn these dreams into reality. China was known to have rockets as early as AD 1232. These "rockets" were arrows propelled by gunpowder. In 1883 Konstantin Tsiolkovsky, a Russian schoolteacher, wrote a detailed theory about the travel of liquid propelled rockets through space. In 1926 Robert Goddard successfully launched the first rocket. It rose only 41 feet into the air but proved that rocket travel was possible. During the 1930's, Germans and Russians began to experiment as well. The Russians launched the first rocket to leave the Earth's atmosphere in 1957.

1. Which country was the first to produce rockets?

_____

2. Name the Russian teacher that first developed the idea of using rockets in space.

_____

3. What made Goddard's 1926 launching a success even though the rocket barely left the ground?

_____

_____

_____

4. Which country was the first to launch a rocket into space? In what year?

_____

_____

5. Look in the encyclopedia to find the name of the first rocket in space.

_____

© 1996 Kelley Wingate Publications   CD-3728

Name _____ Skill: Space Exploration

Space Race

The Space Age began on October 4, 1957 when the Soviet Union launched their rocket, Sputnik, that orbited the Earth for 92 days. Three months later the United States sent Explorer 1 to orbit the Earth. The race for space exploration was on! In October 1959 the Soviet Moon probe, Luna 3, sent back the first pictures of the dark side of the Moon. In 1961 Soviet cosmonaut Yuri Gagarin was the first man to leave the Earth's atmosphere and orbit our planet. An unmanned landing on the moon was accomplished in 1966 with the Soviet craft Luna 9. The US Apollo 8 carried men around the moon in 1968 and Apollo 11 (1969) carried the first men to walk on the surface of the Moon. Manned flights have not gone further than the Moon, but many unmanned probes have been sent close to planets like Saturn, Uranus, and Neptune.

1. **Which two countries were the leaders in early space exploration?**

_____

2. **Write the name of the country that was first in each event listed below:**

   first to put a man in space _____

   first to walk on the Moon _____

   first to orbit the Moon _____

   first to orbit the Earth _____

   first to land on the Moon _____

3. **What was the name of the first man in space?**

_____

4. **Where have unmanned space probes been sent?**

_____

5. **What is the name of the first man to walk on the Moon? (Use an encyclopedia to find the answer.)**

_____

© 1996 Kelley Wingate Publications

Name _____ Skill: Space Exploration

First Astronauts

The first astronauts and cosmonauts (Russian astronauts) were neither men nor women. They were animals! Scientists were not sure what affect space travel would have on humans. In 1948 the United States began to send monkeys up in rockets to see what happened as they left and returned to Earth's atmosphere. The Soviet craft, Sputnik 2, was launched in November 1957 with a dog named Laika aboard. For the next several years dogs and monkeys were the first "astronauts" in space. In 1961 Yuri Gagarin of the Soviet Union became the first man to go outside the Earth's atmosphere. Most of the first astronauts were men selected from the military. The first women were selected for the space shuttle programs, although Russia used women much earlier in their programs.

1. Why were animals the first living beings launched in space?

_____

_____

2. What was the first animal to orbit the Earth? What was her name?

_____

_____

3. What is a cosmonaut?

_____

4. When did U.S. women first become astronauts?

_____

_____

5. How many years were animals used before the first human went into space?

_____ years

© 1996 Kelley Wingate Publications 74 CD-3728

Name _____ Skill: Space Exploration

Early Astronauts

Yuri Gagarin was the first human being to orbit the Earth. He was aboard Sputnik 2, a Soviet spacecraft. Alan Shepard became the first United States astronaut in 1961. He travelled in space only fifteen minutes before he splashed down in the ocean. John Glenn (U.S.) orbited the Earth in 1962. Scientists watched him carefully as he ate a meal to see how food was affected in space. Valentina Tereshkova, a Soviet cosmonaut, became the first woman in space when she flew Vostok 6 around the Earth 48 times in June, 1963. In 1969 Neil Armstrong (U.S.) became the first man to walk on the surface of the Moon. As his foot touched the Moon he said, "That's one small step for man, one giant leap for mankind."

1. Who was the first woman to travel in space?

_____

2. What is the name of the first person to walk on the Moon? When did he do this?

_____

_____

3. Which astronaut was tested eating in space?

_____

4. Why do you think the first astronaut in space only orbited the Earth instead of heading directly to the Moon?

_____

_____

5. What do you think Neil Armstrong meant by his statement?

_____

_____

_____

© 1996 Kelley Wingate Publications   75   CD-3728

Name _____ Skill: Space Exploration

Space Time Line

Choose each event from the box at the bottom of this page and write it over the correct date on this timeline.

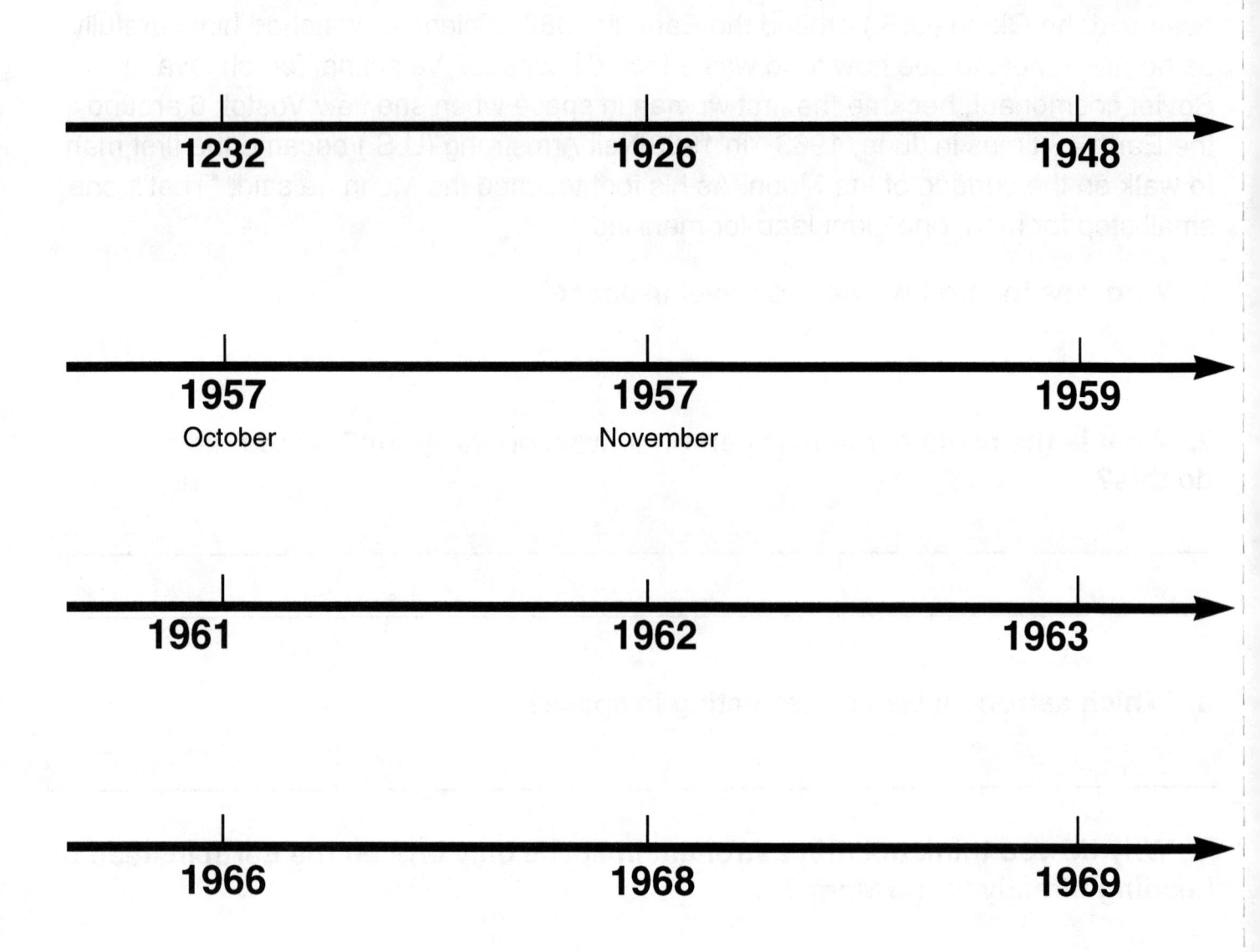

```
|―――――|―――――――――――|―――――――――――|――――→
    1232         1926         1948
```

```
|―――――|―――――――――――|―――――――――――|――――→
    1957         1957         1959
   October     November
```

```
|―――――|―――――――――――|―――――――――――|――――→
    1961         1962         1963
```

```
|―――――|―――――――――――|―――――――――――|――――→
    1966         1968         1969
```

Valentina Tereshkova	**Sputnik**	**Photos: back of Moon**
rocket arrows	Laika	Goddard's rocket
Man walks on Moon	Yuri Gagarin	monkeys in space
Luna 9	John Glenn eats in space	Apollo 8

© 1996 Kelley Wingate Publications   76   CD-3728

Name _____ Skill: Space Exploration

Who Am I?

Match the names to the facts about them.

NAMES

_____ **Neil Armstrong**

_____ **Sputnik**

_____ **Rocket arrows**

_____ **monkeys**

_____ **Soviet Union**

_____ **Robert Goddard**

_____ **Valentina Tereshkova**

_____ **Apollo 11**

_____ **Yuri Gagarin**

_____ **Luna 3**

_____ **Alan Shepard**

_____ **Konstantin Tsiolkovsky**

_____ **John Glenn**

_____ **Laika**

FACTS

A. The Chinese used gunpowder to shoot me at enemies.

B. I wrote about liquid propelled rockets in 1883.

C. I launched the first rocket 41 feet in the air.

D. We put the first man in space.

E. I was the first Soviet rocket in space.

F. I carried the first astronauts to walk on the Moon.

G. We were the first animals to travel in rockets.

H. I was the first dog to orbit the Earth.

I. I was the first man in space.

J. I was the first United States astronaut in space.

K. I was tested eating in space.

L. I was the first woman cosmonaut to travel in space.

M. I was the first person to walk on the Moon.

N. I took the first pictures of the back of the Moon.

© 1996 Kelley Wingate Publications

Name _____ Skill: Space Exploration

The Space Shuttle

The Space Shuttle is the first spacecraft that can be used for more than one flight. It takes off like a rocket, orbits the Earth like a satellite, and lands like an airplane. The Shuttle has four main parts: The orbiter, the rocket boosters, the external fuel tank, and a set of three main engines. Only the external fuel tank must be remade for each new launch. The orbiter has a cargo bay used to carry up to 29,300 kg (65,000 pounds) of payload, the name for space equipment. The external fuel tank carries liquid oxygen and liquid hydrogen. These fuels are burned as the shuttle escapes the Earth's gravity. When the fuel is all burned the tank drops into the ocean. The Shuttle is covered with thousands of insulated silicon tiles. These tiles prevent the Earth's atmosphere from burning up the Shuttle during re-entry.

1. **Name the four parts of the Space Shuttle.**

    1. _____

    2. _____

    3. _____

    4. _____

2. **Which part of the Space Shuttle cannot be reused? Why not?**

_____

_____

_____

_____

3. **What protects the Space Shuttle as it re-enters the Earth's atmosphere?**

_____

_____

_____

© 1996 Kelley Wingate Publications   CD-3728

Name _____ Skill: Space Exploration

The Crew Cabin

The crew cabin of the Space Shuttle is fairly large (71.5 cubic meters or 2,525 cubic feet) with two floors. The air is pressurized so no spacesuits are needed. The top level is the flight deck where the commander and pilot sit. Behind them sit the mission specialists who conduct the experiments on board. The bottom level holds the living quarters where the crew can eat, sleep, and relax. The beds are bunks with a sleeping pallet, pillow, light, and fan. If they prefer, crew members may grab a sleeping bag and hang it on a wall for a nap.

1. How large is the crew cabin?

   _____

2. Why don't astronauts need to wear space suits while in the Shuttle?

   _____

3. What is on the top floor of the crew cabin?

   _____

   _____

   _____

4. Where does an astronaut sleep during a mission?

   _____

   _____

   _____

5. Name the three kinds of people aboard a Shuttle.

   1. _____

   2. _____

   3. _____

© 1996 Kelley Wingate Publications   79   CD-3728

Name _____ Skill: Space Exploration

Spacelab

Spacelab is an international space project for conducting experiments in space. Ten European countries helped develop a laboratory that fits into the cargo bay of the U.S. Space Shuttle. Scientists and others from these countries make up part of the crew called payload (equipment) specialists. From the crew cabin, they enter a tunnel into the laboratory. In the Spacelab they try many experiments to see how things react differently in space. They may grow crystals, test animals, photograph the Earth, or use special telescopes to observe other planets as they orbit the Earth. What a great place to study science!

1. What is Spacelab?

_____

_____

2. Where is Spacelab carried on the Space Shuttle?

_____

_____

3. Who are payload specialists?

_____

4. Name four of the countries that helped develop Spacelab. (Use the encyclopedia to find the answer).

    1. _____

    2. _____

    3. _____

    4. _____

5. What kind of experiment would you like to conduct in the Spacelab?

_____

_____

© 1996 Kelley Wingate Publications   CD-3728

Name _____ Skill: Space Exploration

Space Stations

A space station is like a city in a satellite that orbits the Earth. It is designed for astronauts spending long periods of time in space. It contains a laboratory, workshop, and living quarters. Other spacecraft can come and go from a space station, so it acts as a sort of space age "hotel". The first space station was Salyut 1 launched by the Soviets in 1971. That year, three cosmonauts spent 23 days there. Since then the Soviets built another space station called Mir, where a cosmonaut set a record for staying 439 days in space! A robot spacecraft delivers any needed supplies and mail.

1. What is the difference between a space station and the Space Shuttle?

_____

_____

_____

_____

2. What is the longest time ever spent in space?

_____

3. Why do you think space stations are important in the development of space travel?

_____

_____

_____

4. If you were to live on a space station for the next year, what one item from home would you like to take? Why?

_____

_____

© 1996 Kelley Wingate Publications   CD-3728

Name _____ Skill: Space Exploration

Lift Off!

How can something as large as a Space Shuttle get enough power to push itself into space? Well, lots of fuel is a good guess! The three main engines of the orbiter fire first, building up power. The two solid rocket boosters (on the side of the external fuel tank) ignite and lift off takes place. After about two minutes the solid rockets blaze out. They fall away from the tank and parachutes carry them gently to the ocean. They are recovered and used again on another flight. Eight more minutes pass before the external tank of fuel is burned completely. This tank is also dropped, but it breaks into thousands of pieces before it plunges into the ocean. Finally, two of the main engines push the orbiter into a path around the Earth.

1. Which part of the Shuttle is recovered from the ocean to be reused?

_____

2. Which part of the Shuttle is never recovered?

_____

3. How many fuel tanks are necessary to lift off a Space Shuttle? Name them.

_____

_____

_____

4. How is a Space Shuttle more economical to fly than earlier rockets?

_____

_____

© 1996 Kelley Wingate Publications
CD-3728

Name _____ Skill: Space Exploration

Microgravity

One of the fun yet frustrating things about traveling in space is a lack of gravity. Although there is some gravity present, the body feels weightless and things seem to "float" in the air. Scientists call this microgravity. Microgravity creates some changes in the human body. Blood shifts from the lower to the upper part of the body. This causes eyes to look smaller and forehead wrinkles to disappear while the waist and feet become smaller. In microgravity a person is taller than they are on Earth because gravity is no longer pushing down the discs in the spine! Microgravity presents problems, too. Astronauts usually get the sniffles, become dizzy, and are often nauseated. The heart, bones, and muscles do not have to work as hard and can become weakened after long periods in microgravity. These problems disappear after a few days back on Earth.

1. What is microgravity?

_____

_____

2. What happens to the blood in your body during a space flight?

_____

_____

_____

3. Why is a person taller in space than on Earth?

_____

_____

4. Which side effects of microgravity are most harmful to your health over a long period of time?

_____

_____

_____

© 1996 Kelley Wingate Publications   CD-3728

Name _____ Skill: Space Exploration

Eating in Space

A Space Shuttle contains a galley, or kitchen, for food preparation. The pantry, oven, trays, and hot and cold water are kept in the galley. Astronauts eat three meals each day, selecting from over 100 foods and twenty beverages kept on board! Menus are made and repeated every seven days. There is no refrigerator on board so food must be dehydrated (water is removed and must be replaced before eating), thermostabilized (heated then put in foil pouches), irradiated (preserved by radiation), dried, or natural. Each crew member takes a turn at galley duty. A typical meal takes about 30 minutes to prepare.

1. What is a galley?

_____

_____

2. How many meals does one astronaut eat in one week?

_____

_____

3. Why wouldn't you find fresh milk, butter, or ice in the Shuttle galley?

_____

_____

_____

4. Name five methods of food preparation for travel in space:

   1. _____

   2. _____

   3. _____

   4. _____

   5. _____

© 1996 Kelley Wingate Publications   CD-3728

Name _____ Skill: Space Exploration

Dinner Time!

Eating in space is not exactly easy. The first problem is staying at the table! Astronauts must strap themselves down to keep from floating around the cabin. Food cannot be served on plates or in bowls as it is on Earth. Food must be served in containers that keep it from floating about the cabin. Getting a drink is another problem. It is not possible to pour water or juice into a glass. Microgravity causes liquids to break up into balls that float around until they hit something and spread out. Astronauts must be very careful not to let crumbs or liquids "spill" because they may get into equipment and cause serious problems. They must eat slowly and carefully to avoid spills.

1. Why must astronauts be so careful about spills and crumbs?

_____

_____

2. How do liquids react when poured in microgravity?

_____

_____

3. List three foods you had today that would not be good for space:

1. _____

2. _____

3. _____

© 1996 Kelley Wingate Publications   CD-3728

Name _____ Skill: Space Exploration

Space Age Food

Below are listed some foods commonly found on a Space Shuttle. Rename the foods using space age words from our solar system. For example, instead of ice cream you could have "Polar Snow".

Common Name	Space Age Name
bread	_____
dried apricots	_____
scrambled eggs	_____
pecan cookies	_____
hot dogs	_____
macaroni and cheese	_____
tuna fish	_____
pudding	_____
orange drink	_____
beef and gravy	_____
applesauce	_____
jam or jelly	_____
bananas	_____

© 1996 Kelley Wingate Publications   CD-3728

Name _____ Skill: Space Exploration

Clothing in Space

The Space Shuttle carries two spacesuits designed for any work outside of the pressurized cabin areas. These suits, called EMU (extravehicular mobility unit), allow astronauts to work in the cargo bay, outside the Shuttle, or on the Moon. The EMU has three parts: the liner, the pressure vessel, and the primary life-support system backpack. The liner is like long underwear filled with tubes of water to keep the person cool. The pressure vessel includes the outer jacket, pants, gloves, and helmet. These pieces come in different sizes that can all be attached to make the suit. The primary life-support system backpack (PLSS) is worn on the back and connects to the suit to supply oxygen. The PLSS holds enough oxygen for seven hours of work outside the cabin.

1. When does an astronaut wear an EMU?

_____

_____

2. What do the letters EMU and PLSS mean?

   EMU _____

   PLSS _____

3. What is unusual about the liner or "long johns" of an EMU?

_____

_____

4. How do astronauts make the EMU's fit the different size bodies?

_____

_____

5. What does the PLSS carry?

_____

© 1996 Kelley Wingate Publications   CD-3728

Name _____ Skill: Space Exploration

Keeping Space Clean

On Earth it is easy to take out the trash, but how is that done in space? Trash and garbage quickly breed unhealthy germs that can rapidly spread in a closed area like the Space Shuttle. The cabin must be kept clean to avoid this problem. Food trays are wiped clean with a wet paper cloth. Empty food containers, leftovers, and paper wipes are put into plastic bags that are then tightly sealed. The sealed bags are tossed into a special storage locker that will be emptied when the Shuttle returns to Earth. Astronauts must also keep themselves clean. They put on fresh underwear daily, a clean shirt every three days, and clean pants once a week. The soiled laundry is sealed in plastic bags and stored for return to Earth.

1. Why must trash be stored so carefully on the Space Shuttle?

_____

_____

2. What do astronauts do with left over food and empty containers?

_____

_____

3. Where is garbage placed on board the Space Shuttle?

_____

4. Why do you think waste materials are brought back to Earth?

_____

_____

5. How could waste and laundry become a problem on a space station?

_____

_____

© 1996 Kelley Wingate Publications   CD-3728

Name _____ Skill: Space Exploration

Re-entry

The time has come for the Space Shuttle to return to Earth. Everything that is not needed must go into storage. Cargo bay doors are shut. Seats come out of storage and are bolted to the floor. The crew members put on antigravity suits to help their bodies adjust to the changes during re-entry, then strap themselves in their seats. Engines are used to slow the spacecraft down. The tiles that line the outside of the Shuttle glow red with friction heat as the craft re-enters the Earth's atmosphere. The shuttle slows from 25 times the speed of sound to subsonic speeds within a few minutes. The craft now flies like an airplane as it approaches a runway. After landing, the crew is carefully checked by physicians before they go home.

1. Why must crew members wear special suits for re-entry?

_____

_____

_____

2. What causes the tiles to turn red?

_____

3. Below are listed some events of re-entry. Put a one beside the event that happens first, a two for second, and so on.

_____ Crew members are strapped into their seats.

_____ The Shuttle is flown like a plane.

_____ Engines slow down the craft.

_____ Seats are bolted to the floor.

_____ The Shuttle re-enters the Earth's atmosphere.

_____ Things are put into storage.

_____ Antigravity suits are put on.

© 1996 Kelley Wingate Publications   89   CD-3728

Name _____ Skill: Space Exploration

Astronaut Acronyms

An acronym is a word or term made up of the first letters of many words. For example, NASA is an acronym for <u>N</u>ational <u>A</u>eronautics and <u>S</u>pace <u>A</u>dministration. Below is a list of common acronyms used by astronauts. Match the acronym with the full term.

_____ TDRSS   A. Portable Life-Support System (backpack with oxygen)
_____ ATU   B. Extravehicular Mobility Unit (Spacesuit with life support)
_____ IVA   C. Extravehicular Activity (work outside the crew cabin)
_____ SOMS   D. Shuttle Orbiter Medical System (medical kit with monitors)
_____ EMU   E. Space Transportation System (shuttle, tank, and boosters)
_____ MCC   F. High-Order Assembly Language for Shuttle (computers)
_____ AFB   G. Audio Terminal Unit (communication unit)
_____ EVA   H. Tracking and Data Relay Satellite System
_____ ETI   I. Air Force Base
_____ SRB   J. Kennedy Space Center
_____ HALS   K. Solid Rocket Booster
_____ ELV   L. Intravehicular Activity (work done inside crew cabin)
_____ KSC   M. Extraterrestrial Intelligence (aliens)
_____ STS   N. Expendable Launch Vehicle (a rocket used only once)
_____ PLSS   O. Mission Control Center

Make up five "space sentences" using at least one space acronym in each.

1. _____
   _____
2. _____
   _____
3. _____
   _____
4. _____
   _____
5. _____
   _____

© 1996 Kelley Wingate Publications

Name _____ Skill: Space Exploration

Space Matching

Match the words on the left with the definitions on the right.

_____ EMU   A. Takes off like a rocket and lands like a plane

_____ Goddard   B. Protects the Shuttle from frictional heat

_____ Sputnik   C. A scientific work area in the cargo bay of the Shuttle

_____ Shepard   D. Anything that orbits another object

_____ Glenn   E. A kind of city in space

_____ Armstrong   F. Feeling of weightlessness

_____ Space Shuttle   G. He launched the first liquid-fueled rocket

_____ silicon tiles   H. The equipment on the Space Shuttle

_____ microgravity   I. A United States person trained for space travel

_____ galley   J. Extravehicular Mobility Unit

_____ PLSS   K. The first U.S. astronaut to orbit the Earth

_____ cosmonaut   L. The first man-made satellite to orbit the Earth

_____ astronaut   M. He spent only 15 minutes in space the first time

_____ payload   N. A Soviet trained for space travel

_____ crew cabin   O. The first man to walk on the Moon

_____ Spacelab   P. Contains enough oxygen for seven hours

_____ satellite   Q. Another name for kitchen aboard the Shuttle

_____ space station   R. 71.5 cubic meters (2,525 cubic feet) on two floors

© 1996 Kelley Wingate Publications   91   CD-3728

Name _____ Skill: Space Exploration

SPACE EXPLORATION CROSSWORD

ACROSS

1. 1st person to successfully launch a rocket.
3. The part of a shuttle used to carry the payload.
5. Spacecraft that can be used again and again.
7. Space traveler trained in Russia.
9. First rocket to orbit the Earth.
11. The kitchen of a spacecraft.
12. A satellite "city" that orbits the Earth.

DOWN

1. Last name of the first man in space.
2. A space traveler trained in the United States.
4. First man to walk on the Moon.
6. Extravehicular Mobility Unit.
8. The feeling of weightlessness.
9. An international project used for space experiments.
10. A place to store trash and dirty laundry.

© 1996 Kelley Wingate Publications   CD-3728

Name _____ Skill: Space Exploration

SPACE EXPLORATION

Read the clues below and give the name of what it describes.
Circle the names in the word search. They may go in any direction.

D	L	S	P	A	C	E	L	A	B	C	Y	Y	C	M
Z	G	A	S	P	U	T	N	I	K	T	E	S	O	U
A	O	S	I	P	J	V	Q	V	I	L	K	H	S	T
S	D	D	P	K	L	Q	N	V	L	L	O	U	M	E
T	D	O	Z	A	A	X	A	A	U	M	L	T	O	R
R	A	S	V	C	C	R	G	Q	L	T	S	T	N	E
O	R	G	D	I	G	E	C	X	Q	T	Y	L	A	S
N	D	P	V	O	J	P	S	A	R	T	P	E	U	H
A	D	E	R	A	V	M	L	T	R	U	S	T	T	K
U	W	C	N	A	H	J	K	F	A	G	S	Q	O	O
T	I	I	G	K	X	J	U	M	B	T	O	S	Z	V
M	H	M	S	G	A	G	A	R	I	N	I	B	I	A
C	A	R	M	S	T	R	O	N	G	J	R	O	A	A
V	X	X	P	L	A	S	T	I	C	B	A	G	N	Y
U	N	I	T	E	D	S	T	A	T	E	S	C	H	U

CLUES
1. The first man on the moon. _____
2. A space traveler from the United States. _____
3. The kitchen aboard the shuttle. _____
4. A large storage area for payloads. _____
5. The country that made "rockets" in 1232. _____
6. A space traveler from Russia. _____
7. The first man to orbit the Earth. _____
8. A scientist who launched a rocket 41 feet. _____
9. A cosmonaut that was a dog. _____
10. So little gravity that objects float in the air. _____
11. A place in spacecraft for dirty laundry and trash. _____
12. The country that put the first man into space. _____
13. A reusable spacecraft. _____
14. A labratory that fits into the cargo bay of the space shuttle. _____
15. A kind of city that orbits the Earth. _____
16. The first rocket to orbit the Earth. _____
17. The first woman in space. _____
18. The country that put the first man on the Moon. _____

© 1996 Kelley Wingate Publications   93   CD-3728

Name _____ Skill: Space Exploration

TRUE OR FALSE?

Read each statement carefully. If the statement is true, put a T on the blank. If the statement is false, put a F on the blank.

_____ 1. The Space Shuttle can be reused for other missions.

_____ 2. The external fuel tank is dropped in space where it disappears.

_____ 3. The external tank carries liquid oxygen and liquid nitrogen.

_____ 4. The top floor of the crew cabin contains the flight deck.

_____ 5. Crew members may sleep in a bunk or in a bag on the wall.

_____ 6. The Spacelab can be found in the cockpit.

_____ 7. Spacelab was developed by five countries.

_____ 8. The first space station was called Mir.

_____ 9. The record for staying in space is 493 days.

_____ 10. All external fuel tanks are dropped from the orbiter before it goes into orbit around the Earth.

_____ 11. In space, your waist and feet get smaller.

_____ 12. Microgravity can give you the sniffles and make you sick or dizzy.

_____ 13. Over 100 foods are stored in the galley.

_____ 14. The galley has a refrigerator, but no freezer.

_____ 15. Crumbs or spills may get into the equipment and damage it.

_____ 16. The EMU liner is filled with tubes of water to keep the astronaut cool.

_____ 17. Silicon tiles protect the Space Shuttle from burning up on re-entry.

© 1996 Kelley Wingate Publications

Name _____ Skill: Space Exploration

WHO AM I?

Read each clue and decide who or what is the answer. Write your answer on the line. Answers may be used more than once!

1. The spacecraft that can be reused. _____

2. The person trained for space travel. _____

3. The only part of the Shuttle that cannot be reused. _____

4. We keep the Shuttle from burning up on re-entry. _____

5. Two floors where the crew works and lives. _____

6. We conduct all experiments during the mission. _____

7. The laboratory kept in the cargo bay. _____

8. A "hotel" or city that orbits Earth. _____

9. We hold fuel for lift-off and can be reused. _____

10. I make you feel weightless. _____

11. The place where food is prepared. _____

12. A backpack that holds seven hours of oxygen. _____

13. I hold all garbage and soiled clothes so germs don't spread. _____

14. I am found behind the crew cabin and carry most of the payload, such as the Spacelab. _____

© 1996 Kelley Wingate Publications   95   CD-3728

Name _____ Skill: Space Exploration

ACROSTIC

Read the clues and figure out the name of the astronomer. Write his name in the spaces provided. (Clue number one goes first, two second, and so on.)

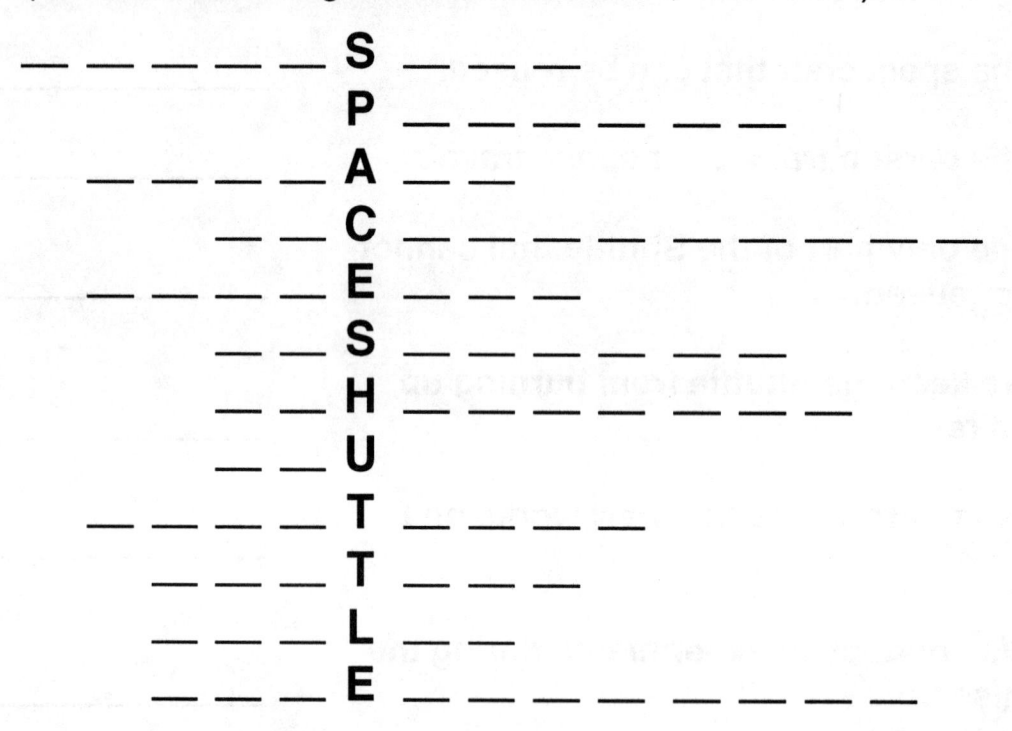

```
    _ _ _ _ S _ _ _
          P _ _ _ _ _ _
      _ _ _ A _ _
        _ _ C _ _ _ _ _ _ _ _ _
      _ _ _ _ E _ _ _
        _ _ S _ _ _ _ _ _
        _ _ H _ _ _ _ _ _ _
          _ _ U
      _ _ _ _ T _ _ _ _
        _ _ _ T _ _ _
        _ _ _ L _ _
        _ _ _ E _ _ _ _ _ _ _
```

1. Clothing, gloves, and helmet worn to protect the space crew.

2. Equipment carried aboard the Space Shuttle.

3. The man who launched the first liquid propelled rocket 41 feet in the air.

4. A feeling of weightlessness.

5. A laboratory built by ten countries and carried in the Space Shuttle.

6. A Soviet astronaut.

7. Food that must have the water replaced before it can be eaten.

8. A spacesuit that is worn for work outside the crew cabin.

9. The first man to walk on the surface of the Moon.

10. The first rocket to orbit the Earth.

11. The kitchen in the Space Shuttle.

12. The only part of the Space Shuttle that must be replaced after it is used.

© 1996 Kelley Wingate Publications
CD-3728

Name _____ Skill: Space Exploration

Crew Patch

Every mission that is taken in space has a crew patch designed to represent it. The patch gives the name of the mission, dates of take-off and re-entry, and sometimes the names of the crew members. As a well known artist who will be one of the crew members on the next mission, you have been selected to design the crew patch. You may pick the planet, the members of your crew, and the mission name. Design your patch then draw and color it below.

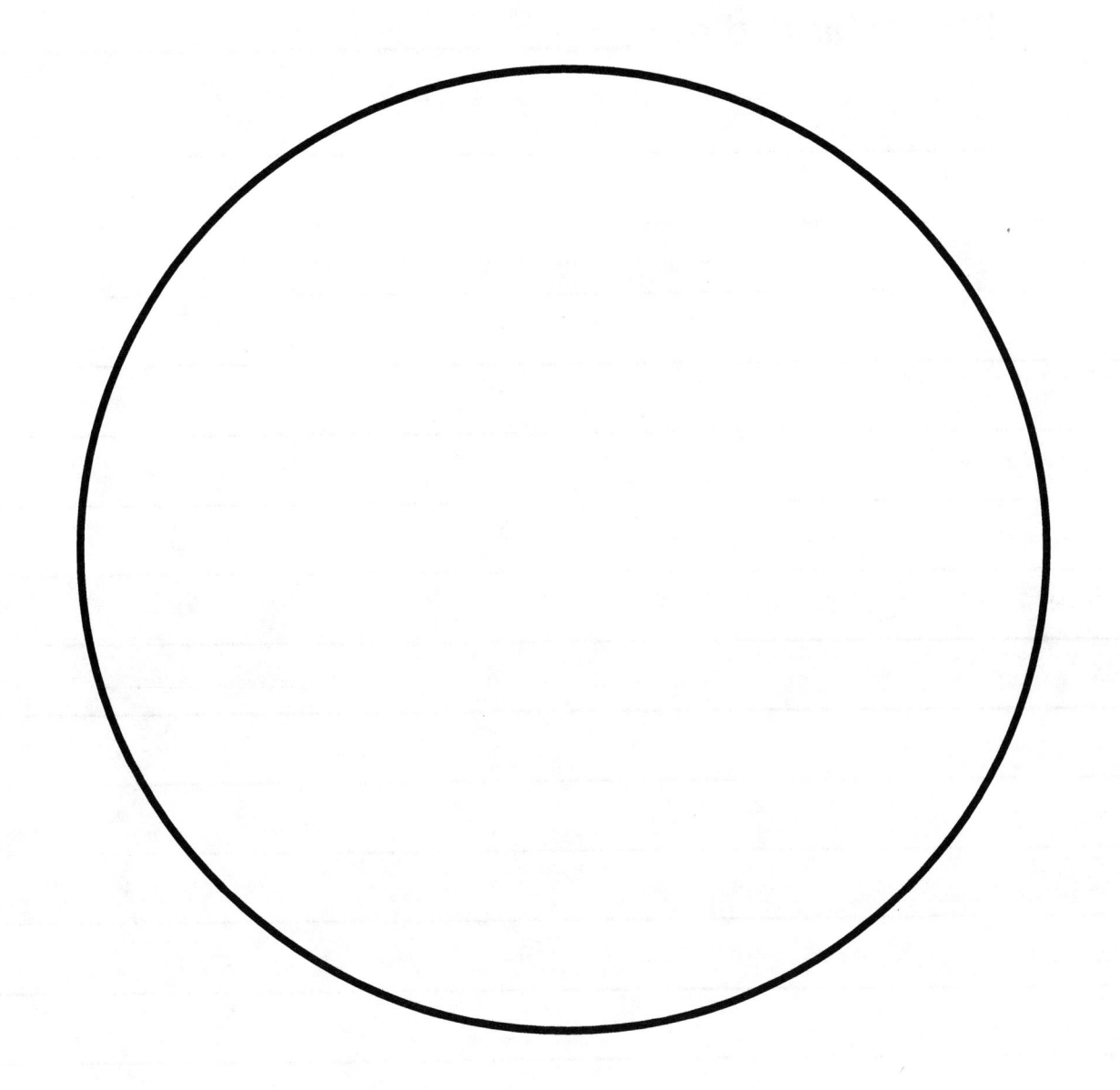

© 1996 Kelley Wingate Publications   97   CD-3728

Name _____ Skill: Space Exploration

Alien: Friend or Foe?

You have been sent from Earth to visit one of the planets in our solar system. Your mission is to search for life on this planet. After traveling through space for months or years, you finally land. As you depart your spaceship you hear strange sounds nearby. Cautiously you move across the surface of the planet. Suddenly something taps your shoulder. You turn and find yourself face to face with an alien.

Write a brief paragraph describing the alien you are facing. Be sure to include things an alien would need to survive on the planet you chose. Then, draw a picture of your alien on the back of this paper.

Planet landed on: _____

_____
_____
_____
_____
_____
_____
_____
_____
_____
_____
_____
_____
_____

© 1996 Kelley Wingate Publications   CD-3728

Name _____ Skill: Recall

How Much Do You Remember?

Finish the statement or answer the question with a word or short answer.

1. A cloud of dust and gas in space is called _____.

2. What is the name of the largest star? _____

3. A star that pulls in its own light is called _____.

4. What did Galileo locate on the Sun? _____

5. What is the age of our Sun? _____

6. A group of stars that make a picture is called _____.

7. Food that is preserved by radiation is called _____.

8. The tubes in the _____ helps keep you cool.

9. _____ takes place when a spacecraft is launched.

10. _____ is the smallest known star.

11. Which astronomer wrote the Almagest? _____

12. 186,000 miles per second is _____.

13. The planet _____ has the most moons.

14. A _____ occurs when the Moon comes directly between the Sun and Earth.

15. What is the distance to the Sun? _____

16. _____ is the planet closest to the Earth.

17. Which planet is known as Earth's sister planet? _____

18. A _____ is like a dirty ball of ice.

© 1996 Kelley Wingate Publications   CD-3728

Name _____ Skill: Recall

1. What comes near the Earth every 76 years? _____

2. _____ is another name for Polaris.

3. About how many food items are carried on the Space Shuttle? _____

4. _____ makes you feel weightless.

5. The astronomer _____ measured the circumference of the Earth.

6. _____ is the name of our galaxy.

7. A large group of "shooting stars" is really a _____ .

8. A gush of fire that rises above the Sun then arches back down is called a _____ .

9. The planet _____ rotates every 24 hours.

10. William Herschel discovered which planet? _____

11. This planet takes 248 years to revolve around the Sun. _____

12. Pieces of rock and metal that orbit the Sun are _____ .

13. _____ is the planet usually furthest from the Sun.

14. The Caloris Basin can be found on _____ .

15. Triton is a moon of _____ .

16. A _____ is a meteor that reaches the Earth.

17. The Great Red Spot is found on the planet _____ .

18. The _____ is a reusable spacecraft.

19. The coolest stars are _____ in color.

© 1996 Kelley Wingate Publications   100   CD-3728

Name _____   Skill: Recall

20. The _____ is a satellite that orbits the Earth.

21. _____ named 48 of the constellations.

22. The astronomer known as the Father of Systematic Astronomy was

23. _____

24. Mercury is _____ kilometers from the Sun.

25. A new star is composed mainly of _____ gas.

26. Galileo found that _____ had phases like the Moon, proving the Sun is the center of the solar system.

27. _____ is also called "The Red Planet".

28. Most stars are the same size as _____ .

29. _____ introduced heliocentric theory.

30. The Great Red Spot is a _____ that is over 300 years old.

31. _____ spent twenty years studying under Plato.

32. Who turned a Dutch toy into a telescope used to observe the stars?
_____

33. The idea that the Sun and all planets revolve around the Earth is called _____ theory.

34. The _____ is the only natural satellite of Earth.

35. Neptune was discovered in what year? _____

36. _____ is the smallest and coldest planet.

37. The astronomer _____ developed the geocentric system.

1996 Kelley Wingate Publications   CD-3728

Name _____ Skill: Recall

56. _____ discovered a nova and developed accurate instruments for measuring movement of stars and planets.

57. _____ theory states that the planets revolve around the Sun.

58. _____ gas gives Neptune its blue color.

59. _____ has two wind systems blowing in opposite directions.

60. The moon called _____ is almost as large as the planet it orbits.

61. The hottest stars are _____ in color.

62. _____ developed three laws of planetary motion.

63. The fifth planet from the Sun is _____ .

64. Our Sun is a _____ star.

65. _____ has more satellites than any other planet.

66. Newton explained the theories of _____ and centrifugal force.

67. Name the four main parts of the Space Shuttle.

1. _____
2. _____
3. _____
4. _____

68. Miranda, Ariel, Umbriel, Titania, and Oberon are the five largest moons of _____ .

© 1996 Kelley Wingate Publications   102   CD-3728

Name _____   Skill: Recall

69. The highest point on Earth is _____ .

70. Jupiter has _____ rings.

71. _____ push the tail of the comet away from the Sun.

72. The star nearest the Earth is _____ .

73. Ursa Major is a constellation with _____ stars.

74. The Asteroid Belt orbits around the Sun between the planets Jupiter and _____ .

75. Solar flares give off _____ that can affect radio waves on Earth.

76. The _____ is a cloud of gas that surrounds the nucleus of a comet.

77. Sunspots have an _____ year cycle.

78. There is a total of _____ different constellations.

79. _____ are bits of dust and rock floating in space, probably left by a passing comet.

80. _____ had rocket arrows in AD 1232.

81. The Big Dipper or Big Bear is also called _____ .

82. An old star that suddenly expands and grows very bright is called a _____ .

83. Asteroids hit the Earth during the _____ , but not since that time.

84. A star usually lives for about _____ years.

85. The tail of Halley's comet is about as long as the distance from the Earth to _____ .

© 1996 Kelley Wingate Publications   103   CD-3728

Super Achiever

_____

receives this award for

_____

Keep up the great work!

_____ _____
signed   date

Great Job!

_____

receives this award for

_____

Great Job!

_____ _____
signed   date

Certificate of Completion

_____

receives this award for outstanding performance in

_____

Congratulations!

_____ _____
signed   date

Solar System Award

_____

receives this award for

_____

You are terrific!

_____ _____
signed   date

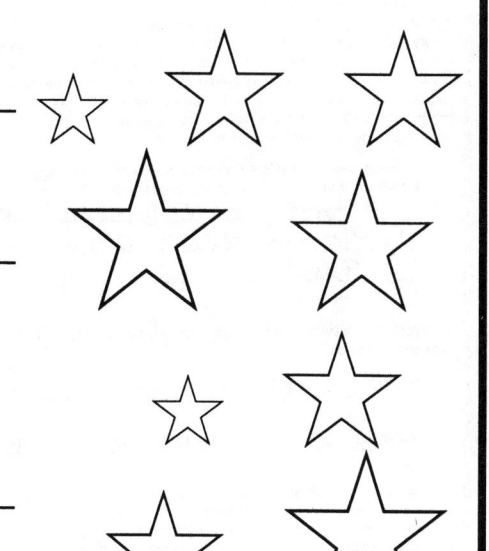

©1996 Kelley Wingate Publications   CD-3728

Answer Key

Pythagoras

Pythagoras was one of the earliest known astronomers. He was born in Greece about 560 B.C. He and a group of other Greeks moved to Croton in southern Italy and set up a school for philosophical and religious ideas. Pythagoras loved mathematics and science, which he applied to the study of the stars and planets. His group believed that the earth revolved around a massive fire contained deep within its core. They also believed that the sun and other planets revolved around this fire.

1. Draw a cross section of the Earth as the Pythagorians thought it might look.

(Drawing of Earth with "Fire" at center)

2. What was the contribution that Pythagoras made to our understanding of the solar system?

He believed the Earth was filled with fire. The planets and Sun revolved around the fire in the Earth.

3. Pretend that we have no information about the solar system. Think of two ways you might learn about the stars and planets. List your methods here.

1. Observe the sky at night
2. Watch the Sun's movement across the sky. (Answers will vary)

Eudoxus of Cnidus

Eudoxus of Cnidus was born in Greece around 408 B.C. He was a scholar interested in the motions of celestial bodies (stars, planets, and moons). Eudoxus was the first person to develop the theory that the stars and planets are spheres that move in concentric circles (evenly spaced within each other) around the earth. This theory is called geocentric, or an earth centered system. This theory helped explain why the sun and stars move across the sky from east to west.

1. What did Eudoxus contribute to our understanding of the solar system?

He added the theory of concentric circles

2. Look up Eudoxus in the encyclopedia. What other subjects did he study?

Math, geography, philosophy, law

3. Draw a simple diagram of the solar system as Eudoxus believed it worked.

(Drawing showing planet orbiting Earth with Sun nearby)

4. "Geo" means earth and "Centric" means centered. Why is the word geocentric a good name for the theory developed by Eudoxus?

The Earth was the center of our solar system in his theory.

Aristotle

Aristotle, a famous Greek scientist, was born about 384 B.C. When he was seventeen he went to Athens to study under Plato, a famous philosopher and one of the greatest thinkers of all time. Aristotle spent twenty years a Plato's school. During that time he began to observe nature, record what he found, and form theories about how nature worked. He organized information and explained many ideas in studies such as biology, ethics, politics, and astronomy. Aristotle agreed with the geocentric theory of Pythagoras and Eudoxus, believing that the planets revolve around the earth. However, he did not believe that the planets move in large circles. His observations led him to believe that the planets move in elliptical, or oval-shaped, orbits. Aristotle was respected as a thinker and so his theory was accepted as fact. The world continued to believe that the earth was the center of the universe for several hundred years until other scientists decided to question this theory.

1. The world believed Aristotle's astrology findings for many years. What are two of those findings presented in this paragraph?

1. Planets move in elliptical orbits.
2. Planets revolve around the Earth.

2. Which of Aristotle's findings do we still believe today? Which "fact" has proven to be false?

True: *Planets move in elliptical orbits*

False: *Planets revolve around the Earth.*

3. Make your own elliptical orbit. Place two push pins or thumbtacks about 10 centimeters apart on a piece of cardboard. Loop a piece of string (about 20 cm long) around the tacks and tie it tightly. Put your pencil inside the loop and, pulling tightly against the string, trace the path of the string. As the string gives somewhat, the pattern will widen in the center, creating a type of elliptical orbit on paper.

Eratosthenes

Eratosthenes was born about 276 B.C in Cyrene, Greece and became the head librarian in the city of Alexandria, Egypt. Eratosthenes made a major contribution to astronomy by calculating the circumference of the earth. At noon during the summer solstice the sun reflected at the bottom of a well in Syene (modern Aswan). Eratosthenes concluded that this city must be in a direct line between the sun and the center of the earth. He calculated the angle of the sun's shadow in Alexandria at exactly noon on the same day. He found that Alexandria was south of Syene by 1/50 of a circle. He measured the distance between the two cities and multiplied that by 50. Eratosthenes calculated the circumference of the earth to be about 39,000 km (or 24,000 miles). He was very close to the actual circumference as we know it today.

1. Use a world atlas to find the actual circumference of the earth. Record the measurement here.

40,075 kilometers
24,902 miles

2. Calculate the difference between the actual circumference and the measurement Eratosthenes found.

1,075 kilometers
902 miles

3. What led Eratosthenes to the conclusion that Syene was in a direct line between the sun and the center of the earth?

The Sun cast no shadow in the well at noon of the summer solstice.

Answer Key

Hipparchus

Hipparchus was born in Nicaea (now Iznik, Turkey) about 190 B.C. He was one of the greatest early astronomers and is known as the Father of Systematic Astronomy. Hipparchus believed that the earth was the center of the universe and all other celestial bodies moved around it in perfect circles. He catalogued over 850 stars and kept a record of their small but important yearly movements in our sky. Hipparchus found better ways of determining the distances and diameters of the Sun and Moon. He was one of the first people to use mathematics in determining longitude and latitude of positions on earth.

1. Look in an encyclopedia to find the distance from the Earth to the Sun and the Moon. Find the diameters of the Sun and Moon.

   Earth to Sun: **150 million** km. **93 million** mi.
   Earth to Moon: **385,000** km. **239,000** mi.
   Diameter of Sun: **1,400,000** km. **865,000** mi.
   Diameter of Moon: **3,476** km. **2,160** mi.

2. Compare the solar system theory of each of these three men.

EUDOXUS	ARISTOTLE	HIPPARCHUS
Center of Universe: **Earth**	Center of Universe: **Earth**	Center of Universe: **Earth**
How planets move: **concentric circles**	How planets move: **elliptical circles**	How planets move: **in perfect circles**

Ptolemy

Ptolemy was born in Greece about 100 A.D. He made many contributions to geography, mathematics, and astronomy. He spent fourteen years in Alexandria, Egypt observing the stars and planets. Ptolemy wrote Almagest, a book that gave a detailed theory involving motion the of the Sun, Moon, and planets. He based his theory on the beliefs of Hipparchus (all celestial bodies revolve around the earth) and Aristotle (planets move in elliptical orbits). Ptolemy added the idea that planets move in epicycles. He believed that the planets were turn around their own centers as they revolve around the earth. His theory was not much improved upon for more than 1000 years.

1. What was the major contribution that Ptolemy gave to the science of astronomy?

   Ptolemy added the idea that planets have epicycles.

2. Ptolemy studied the science of cosmology. What is cosmology? Use an encyclopedia or dictionary to find the answer.

   Cosmology is the study of the universe.

3. Ptolemy later wrote a book about the earth and called it Geography. What did he attempt to do in this book? Use an encyclopedia to find the answer.

   Ptolemy tried to map the known world using longitudes and latitudes.

Copernicus

Nicalous Copernicus was born in Poland on February 19, 1473. He studied mathematics, astronomy, medicine, and law. He practiced medicine and law for his church for many years, but he never lost his love for astronomy. In 1514, at the age of 40, Copernicus wrote a paper challenging the geocentric theory. He agreed with Hipparchus that the planets revolve in perfect circles. He also agreed with Ptolemy that the planets have epicycles, spinning in smaller orbits as they make the larger revolution. However, Copernicus introduced the idea that the earth and planets revolve around the sun. This new idea became known as the heliocentric (sun centered) theory.

1. What was Copernicus' major contribution to the study of astronomy?

   Copernicus believed the Sun was the center of the solar system.

2. In what ways did Copernicus agree with Hipparchus and Ptolemy?

   Copernicus believed in concentric circles as orbits like Hipparchus.

   Copernicus believed in epicycles like Ptolemy.

3. Look up geocentric and heliocentric in the dictionary. What does each word mean?

   GEOCENTRIC: **Earth-centered solar system**
   HELIOCENTRIC: **Sun-centered solar system**

Brahe

Tycho Brahe was born December 14, 1546 in Denmark. He studied law, but became interested in astronomy when he observed a solar eclipse in 1560. He read Ptolemy's Almagest and went on to study science in several universities. In 1572 Brahe discovered a nova (a star that becomes very bright then fades in a few months or years) and five comets that were beyond the Moon's orbit. His findings did not agree with either Ptolemy or Copernicus so Brahe developed his own theory called the Tychonic system. This system suggests that the Sun and Moon revolve around the Earth and all other planets revolve around the Sun. Although his theory is not accurate, Brahe did contribute a great deal to the science of astronomy. He developed new and more accurate instruments for measuring the movement of stars and planets. He also compiled a large collection of observations that became useful to his student and future astronomer, Kepler.

1. What was the major contribution Brahe gave to astronomy?

   Brahe developed instruments for more accurate measurement of stars and planets

2. Explain, in your own words, what a nova and a comet are.

   NOVA: **is a star that becomes bright and then fades. (Answers will vary.)**

   COMET: **is a ball of ice, rock, dust. (Answers will vary.)**

3. On the back of this paper, draw a diagram of how Brahe thought the universe must look.

© 1996 Kelley Wingate Publications — CD-3728

Answer Key

Galileo

Galileo Galilei was born in Italy on February 15, 1564. As a young boy Galileo hated science because he did not believe what was taught. He felt theories needed to be proved before they could be believed. Galileo discovered that mathematics could provide a way to test (and prove or disprove) scientific theories. In 1604 Galileo became interested in astronomy when a Supernova appeared. In 1609 he was introduced to a Dutch toy that made far away things look closer - the first telescope. Galileo designed and built his own much larger telescope and turned it toward the sky. He observed the Moon and several "planets" circling Jupiter (he later identified these as moons). Galileo noticed that Venus had phases just like the moon. This observation gave him the first proof that the planets did indeed revolve around the Sun as Copernicus had stated.

1. Why didn't Galileo care much for science? What changed his mind?

   Galileo did not believe in many science "facts". He wanted proof.

2. In your opinion, which was Galileo's greater discovery - the use of the telescope or proof that the planets circled the Sun? Justify your answer with facts.

   (Answers will vary.)

3. Did Galileo's findings support a geocentric or heliocentric theory?

   Galileo's findings support a heliocentric theory.

Kepler

Johannes Kepler was born on December 27, 1571 in Germany. He studied astronomy under Tycho Brahe and supported most of the theories developed by Copernicus. Kepler observed the orbit of planets and was the first to accurately determine their position and how they orbited the sun. Kepler developed three laws regarding planetary motion. The first law states that the sun is the center of the universe and that the planets revolve around it in elliptical orbits. The second law states that planets move faster when their orbits bring them closer to the sun. The third law is a mathematical formula used to determine the distance of a planet from the sun.

1. Kepler proved the planets orbited the sun in elliptical orbits. How did this agree with the teachings of Copernicus? How did this disagree with Copernicus?

   AGREE: The planets orbit the Sun.

   DISAGREE: Copernicus believed in concentric circles.

2. Kepler wrote six books during his lifetime. Look in the encyclopedia and find the names of two of his books and the date they were published.

   1596 - Cosmographic Mystery 1618 - Intro to Coper. Astron.
   1609 - New Astronomy 1627 - Rudolphic Tables
   1619 - Harmonies of the World 1604 - Theory of Orbits

3. What was the major contribution that Kepler made to our understanding of the solar system?

   Three laws to explain planet movements.

Newton

Sir Isaac Newton was born on January 4, 1643 in England. He is often noted as the greatest scientific genius of all time because he made important contributions to every major area of science known during his time: Mathematics, physics, optics, and astronomy. In the field of astronomy, Newton is best known for his laws of motion and gravity. People already knew that objects dropped would fall to the ground, but they had no idea why that happened. Newton explained that there was a force called gravity that pulled objects toward the center of the earth. Newton developed the theory of gravitational pull. Gravity is a force that pulls objects toward the center of an area. The Earth has a gravitational pull that attracts even the moon! The sun also has gravity that tries to pull the Earth and other planets toward its center. Newton explained a law of motion called centrifugal force that stops the planets from being pulled into the sun. As an object revolves in a circular path, centrifugal force causes the object to pull away from the center of the orbit. A balance between gravity and centrifugal force holds the planets in a steady orbit around the sun.

1. What important contribution did Newton make to the science of astronomy?

   Newton added the laws of motion and gravity.

2. According to Newton, why does a planet orbit around the sun?

   Gravity pulls it toward the Sun. Centrifugal force pulls it away from the Sun.

3. Newton wrote a book that is considered to be the greatest scientific book ever written. What is the name of that book? (Use an encyclopedia to help you find the answer.)

   Principia - (The Mathematical Principles of Natural Philosophy)

Herschel

Sir William Herschel was born in Germany on November 15, 1738. He was a music teacher but spent his spare time studying astronomy and building telescopes. He constructed the largest reflecting telescopes of his time using them to observe the six known planets: Mars, Mercury, Earth, Venus, Jupiter, and Saturn. In 1781, Herschel looked beyond the known solar system and made an amazing discovery - a new planet! He named this seventh planet "Georgium Sidus" after England's King George III. In 1782 Herschel constructed an even larger telescope and found Georgium Sidus has two moons. He also discovered two new moons of Saturn. The name Herschel gave his planet never really stuck. Astronomers preferred to call the planet "Herschel", but by the mid 1800's it was commonly known by a name taken from mythology.

1. What was Sir William Herschel's contribution to the science of astronomy?

   Herschel discovered a new planet.

2. Using the encyclopedia, look up Sir William Herschel and find the name that we commonly use for the planet he called Georgium Sidus.

   Uranus

3. Pretend that you have just discovered a new planet in our solar system. What would you name the planet? Explain why you chose that name.

   (Answers will vary.)

Answer Key

Page 13

Name _____ Skill: Theory

Matching Astronomers
Match the astronomer to the fact about him.

ASTRONOMER		FACT
D	Aristotle	A. He discovered the seventh planet in 1781.
E	Brahe	B. He wrote the Almagest
L	Copernicus	C. He developed the theory that planets are spheres and move in concentric circles around the Earth.
K	Eratosthenes	
C	Eudoxus	D. He introduced the idea of elliptical orbits in geocentric theory.
F	Galileo	
H	Hipparchus	E. He developed the Tychonic system and was the first to discover a nova.
A	Herschel	
J	Kepler	F. He used his telescope to prove the heliocentric theory.
G	Newton	G. He explained gravity and centrifugal force.
B	Ptolemy	
I	Pythagoras	H. He used math to determine longitudes and latitudes of the Earth.
		I. He believed the planets revolve around a fire at the core of the Earth.
		J. He wrote three famous laws about the movement of planets.
		K. He measured the distance around the Earth.
		L. The first astronomer to name the sun as the center of the solar system.

Page 14

Name _____ Skill: Theory

Matching Astronomers
Match the astronomer to the fact about him.

ASTRONOMER		FACT
D	Aristotle	A. He challenged the geocentric theory in 1514.
K	Brahe	B. He discovered Georgium Sidus.
A	Copernicus	C. He developed the theory that planets move in concentric circles around the Earth.
J	Eratosthenes	
C	Eudoxus	D. He studied under Plato for twenty years.
G	Galileo	E. He was called the Father of Systematic Astronomy.
E	Hipparchus	
B	Herschel	F. He set up a school for philosophy and religion in Croton, Italy.
L	Kepler	G. He proved heliocentric theory by discovering the phases of Venus.
H	Newton	
I	Ptolemy	H. He is called the greatest scientific genius of all time.
F	Pythagoras	I. He believed that the planets move in epicycles.
		J. He was the head librarian in Alexandria, Egypt.
		K. He was the first to discover a nova (1572).
		L. He studied under Tycho Brahe.

Page 15

Name _____ Skill: Theory

Read the clues and figure out the name of the astronomer. Write his name in the spaces provided. (Clue number one goes first and begins with an A.)

```
       A R I S T O T L E
     H E R S C H E L
         N E W T O N
  P Y T H A G O R A S
       G A L I L E O
E R A T O S T H E N E S
           C O P E R N I C U S
         P T O L E M Y
       K E P L E R
         B R A H E
```

1. He studied under another astronomer named Plato.
2. He named a new planet "Georgium Sidus".
3. He is noted as the greatest scientific genius of all time.
4. He believed that there was a huge fire inside the Earth.
5. He turned a Dutch toy into an instrument to view the sky more clearly.
6. He measured the sun's angle between two cities to find the circumference of the Earth.
7. He introduced the heliocentric theory.
8. He added the theory of epicycles to the study of astronomy.
9. He developed a mathematical formula to determine the distance between the planets and the sun.
10. He believed the Sun and Moon revolve around the Earth, but all other planets revolve around the Sun.

Page 16

Name _____ Skill: Theory

Astronomer Chart

Look at the list of astronomers below the chart. List them under the proper heading according to which system they believed in - geocentric or heliocentric.

GEOCENTRIC	HELIOCENTRIC
Eudoxus	Copernicus
Aristotle	Galileo
Ptolemy	Kepler
Hipparchus	Herschel
Pythagoras	Newton

Copernicus Aristotle Hipparchus
Galileo Ptolemy Pythagoras
Eudoxus Kepler Newton
　　　　　Herschel

Answer Key

Page 17 — Astronomer Crossword (Skill: Theory)

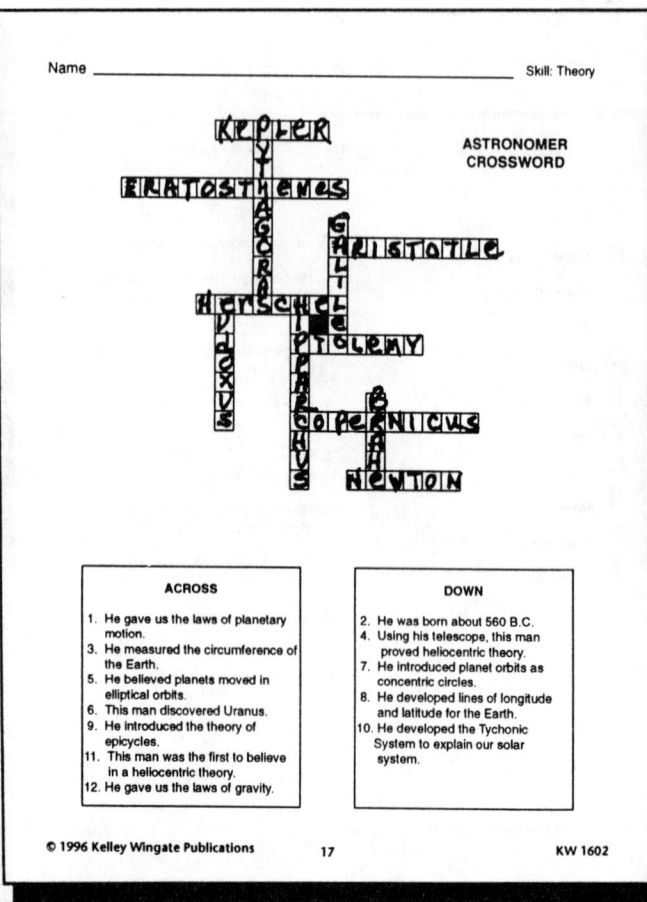

Across:
1. KEPLER — He gave us the laws of planetary motion.
3. ERATOSTHENES — He measured the circumference of the Earth.
5. ARISTOTLE — He believed planets moved in elliptical orbits.
6. HERSCHEL — This man discovered Uranus.
9. PTOLEMY — He introduced the theory of epicycles.
11. COPERNICUS — This man was the first to believe in a heliocentric theory.
12. NEWTON — He gave us the laws of gravity.

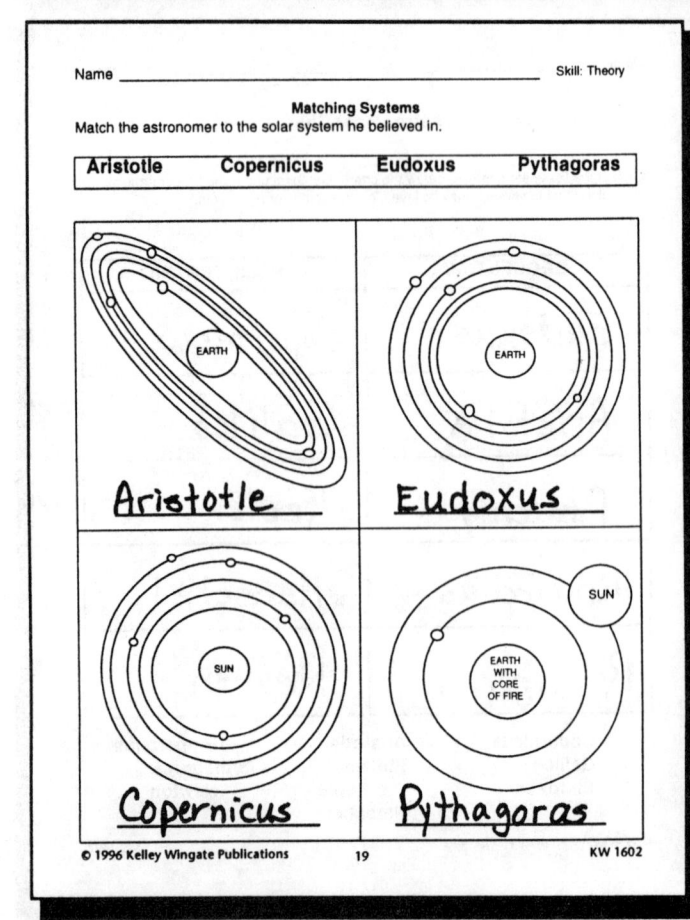

Down:
2. PYTHAGORAS — He was born about 560 B.C.
4. GALILEO — Using his telescope, this man proved heliocentric theory.
7. EUDOXUS — He introduced planet orbits as concentric circles.
8. HIPPARCHUS — He developed lines of longitude and latitude for the Earth.
10. BRAHE — He developed the Tychonic System to explain our solar system.

Page 18 — Astronomers Word Search (Skill: Theory)

Word list:
ARISTOTLE, ASTRONOMY, BRAHE, COPERNICUS, ELLIPTICAL, EPICYCLE, ERATOSTHENES, EUDOXUS, GALILEO, GEOCENTRIC, HELIOCENTRIC, HERSCHEL, HIPPARCHUS, KEPLER, NEWTON, NOVA, ORBIT, PLANET, PYTHAGORAS, SOLAR, TELESCOPE

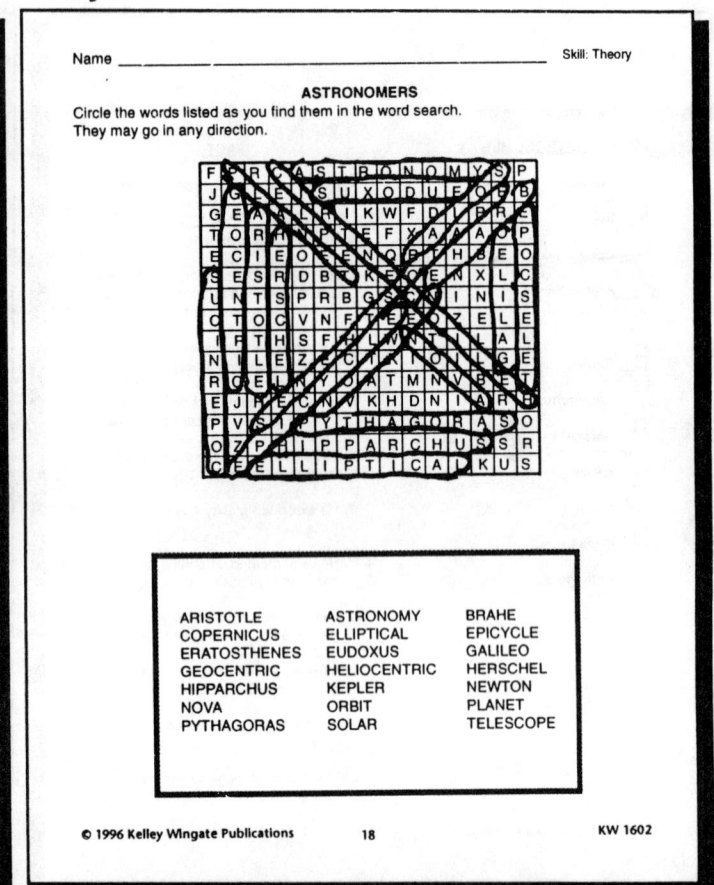

Page 19 — Matching Systems (Skill: Theory)

- Top left (elliptical orbits around Earth): Aristotle
- Top right (concentric circles around Earth): Eudoxus
- Bottom left (planets around Sun): Copernicus
- Bottom right (Sun orbits Earth with core of fire): Pythagoras

Page 20 — Astronomer Time Line (Skill: Space Exploration)

Astronomer	Year
Pythagoras	560 BC
Eudoxus	408 BC
Aristotle	384 BC
Erathosthenes	276 BC
Hipparchus	190 BC
Ptolemy	100 AD
Copernicus	1473
Brahe	1546
Galileo	1564
Kepler	1571
Newton	1643
Herschel	1738

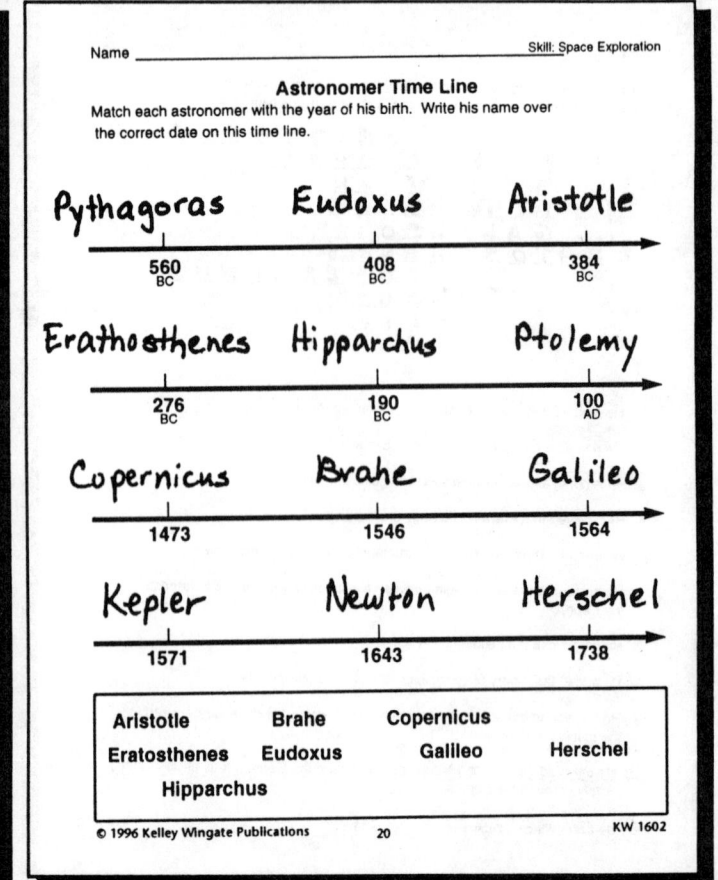

Answer Key

Mercury

Mercury is the closest planet to our Sun. It is a small planet, only a little larger than our moon. Mercury is one of the five planets that can be seen with the naked eye, but it is the most difficult to see. Mercury rotates slowly. One day on Mercury is equal to 59 Earth days! The atmosphere on Mercury is very thin, allowing a lot of heat to be absorbed during the day. At night the atmosphere cannot hold the heat so it becomes extremely cold very quickly. Mercury is covered with craters made by meteors crashing into it. One large area of craters is called Caloris Basin. Mercury's orbit is unusual in that it is more elliptical (oval shaped) than most other planets.

Fact Box
- Distance from Sun: 57.8 million kilometers (36 million miles)
- Rotation: 59 Earth days
- Revolution: 88 Earth days
- Diameter: 4,880 kilometers (3,030 miles)
- Gravity: about 1/3 of the Earth's gravity
- Orbital Speed: 48 kilometers per second (30 miles per second)
- Atmosphere: extremely thin
- Temperature: 350° C (660° F) day -170° C (-270° F) at night
- Rings: none
- Satellites: none
- Travel time from Earth: Jet: 10 years, 8 months Rocket: 3 months Light Years: 5 minutes
- Named for: Roman god, Mercury (Greek god, Apollo)

1. If you were staying on Mercury for one week, how long would that be in Earth days?
   59 × 7 = 413 Earth days

2. Why is Mercury's temperature so hot during the day then so cold at night?
   The atmosphere is very thin. This allows the Sun to heat the planet during the day and heat to escape at night.

3. What does the surface of Mercury look like? Describe it in your own words.
   It is covered with craters.
   Answers will vary.

Venus

Venus is sometimes called the sister planet to Earth because they are so close in size. However, Venus rotates the opposite of the Earth, with the sun rising in the west and setting in the east. The atmosphere of Venus is very thick. If you were to stand on that planet the air would feel like 3,000 feet of water crushing you! Venus has a thick cloud layer that helps reflect the sun so it looks very bright in the night sky. Those clouds also hold the sun's heat next to the surface of the planet, making it extremely hot. The surface of Venus is similar to Earth. It has mountains, valleys, plains, continents, and even areas where oceans could be (if there were any water). Venus also has volcanoes, some of which are many miles wide.

Fact Box
- Distance from Sun: 108 million kilometers (67 million miles)
- Rotation: 243 Earth days
- Revolution: 225 Earth days
- Diameter: 12,100 kilometers (7,520 miles)
- Gravity: about 9/10 of the Earth's gravity
- Orbital Speed: 35 kilometers per second (22 miles per second)
- Atmosphere: 98% carbon dioxide
- Temperature: 400° C (800° F)
- Rings: none
- Satellites: none
- Travel time from Earth: Jet: 5 years, 5 months Rocket: 1.5 months Light Years: 2.5 minutes
- Named for: Roman goddess Venus (Greek goddess Aphrodite)

1. Why does Venus shine so brightly in the night sky?
   The thick cloud layer reflects the Sun's light.

2. Name six ways in which Venus and the Earth are alike:
   1. mountains 2. valleys
   3. plains 4. continents
   5. places for oceans 6. size

3. If you visited Venus for one week, how many Earth days would that be? How many Earth years?
   Seven Venus days equal 243 × 7 = 1,729 (1,701) Earth days.
   Seven Venus days equal 4.7 Earth years.
   1729 ÷ 365 = 4.7

Earth

Earth is the third planet from the Sun and the only known planet to have life. The Earth looks very beautiful from space. It is mostly blue (three-fourths of its surface is covered with water), with brown and green continents and white ice-capped poles. At any given time about half of the Earth is covered with clouds which protect the surface from extreme heat. The clouds also retain, or hold, the heat that is absorbed so that the planet does not become too cold as it rotates away from the sun. The Earth is tilted on its axis, rotating at an angle which allows us to experience seasons. The highest point on this planet is Mount Everest at 5.5 miles above sea level. The lowest point is found in the Dead Sea and is 395 meters (1,296 ft) below sea level.

Fact Box
- Distance from Sun: 150 million kilometers (93 million miles)
- Rotation: 23 hrs 56 min
- Revolution: 365.3 days
- Diameter: 12,800 kilometers (7,900 miles)
- Gravity:
- Orbital Speed: 24 kilometers per second (15 miles per sec.)
- Atmosphere: 78% nitrogen, 21% oxygen, 1% other gases
- Temperature: varies with location
- Rings: none
- Satellites: one (the Moon)
- Travel time from Earth: Jet: Rocket: Light Years:
- Named for: Greek goddess Gaea, mother of the Titans

1. What makes the Earth different from every other planet in our solar system?
   It has life on it.

2. How do clouds help protect our planet?
   Clouds protect the Earth from extreme heat. They also keep the heat from escaping at night.

3. What feature of the Earth gives it seasons?
   The Earth is tilted on its axis, giving us the seasons as it orbits the Sun.

The Moon

The only natural satellite of the planet Earth is the Moon. It is about one fourth the size of the Earth and has no air, food, or water. The gravity of the Moon is too weak to hold an atmosphere although it is strong enough to cause ocean tides on the Earth. The Moon rotates at the same rate it revolves, so the same side always faces the Earth. Humans did not know what was on the dark side of the Moon until 1959 when a Russian spacecraft brought back pictures. In 1969, astronauts walked on the surface of the Moon for the first time. About three fourths of the Moon is covered with mountains and craters formed by crashing meteors. The craters are as large as 1000 km. (620 mi.) across to as small as dots. The rest of the Moon is smooth and dark, commonly called "seas".

Fact Box
- Distance from Earth: 385,000 kilometers (239,000 miles)
- Rotation: 27 days, 7 hrs, 43 mins
- Revolution: 27 days, 7 hrs, 43 mins
- Diameter: 3,476 kilometers (2160 miles)
- Gravity: 1/6 of Earth
- Orbital Speed: 1 kilometer per second (6 miles per second)
- Atmosphere: none
- Temperature: 1300 C (2660 F) day to -1730 C (-2800 F) at night
- Rings: none
- Satellites: none
- Travel time from Earth: Jet: 4 months Rocket: 3 days Light Years: 1.2 seconds

1. Why does the same side of the Moon always face the Earth?
   The moon rotates at the same rate as it revolves.

2. What caused all of the craters on the Moon?
   Crashing meteors caused the moon's craters.

3. List three facts about the Moon's gravity:
   1. It is too weak to hold an atmosphere.
   2. It causes tides on the Earth.
   3. It is 1/6 that of Earth's gravity.

4. Why does the Moon get so hot during the day and so cold at night?
   No atmosphere to hold the heat.

Answer Key

Mars

Mars is the fourth planet from the Sun. It is a rocky planet that is much smaller and colder than Earth. The atmosphere of Mars is too cool for water to exist, but the two poles are covered with a thin layer of ice. Mars is also called "the Red Planet" because a large amount of iron in the soil gives it a reddish color. An interesting feature of this planet is a gigantic canyon named Valles Marineris that stretches for 4,000 km (2,500 mi.). During the 1700's, astronomers observed swirls and dark patches thought to be clouds and seas. It was believed that Mars had an atmosphere that could support life. Many fictional stories centered on the possibility of "Martians." In recent years several space probes have collected information about Mars that proves no life could exist on the surface of that planet.

Fact Box
Distance from Sun: 227 million kilometers (140 million mi.)
Rotation: 24 hrs 40 mins
Revolution: 687 Earth days
Diameter: 6,800 kilometers (4,200 miles)
Gravity: about 1/3 of the Earth's gravity
Orbital Speed: 24 kilometers per second (15 miles per second)
Atmosphere: 95% carbon dioxide, 3% nitrogen, 2% argon
Temperature: -53° C (64° F) to -128° C (-199° F)
Rings: none
Satellites: two
Travel time from Earth:
Jet: 8 years, 10 months
Rocket: 2.5 months
Light Years: 4 minutes
Named for: Roman god, Mars (Greek god Ares)

1. Why is Mars often called the Red Planet?
   The soil is full of iron.

2. What observations led earlier astronomers to believe Mars might support life?
   Early astronomers thought there were clouds and seas.

3. How do we know so much about the surface of Mars when no man has ever set foot on the planet?
   Space probes have collected information about the planet Mars.

Jupiter

Jupiter, the fifth planet from the Sun, is the first of the four gas planets. Jupiter is so large that 2/3 of all the planets in our solar system could fit inside it! The planet is covered with three thick layers of clouds that extend 30 km (19 mi.) above the surface. There is no solid surface on the planet. Jupiter is composed of gases that become liquids about 1/4 of the way to the core (center). Because it rotates so quickly, Jupiter appears flattened around the center and longer at the poles. An interesting feature of Jupiter is a number of large white spots that space probes have identified as storms. There is also the Great Red Spot which is as large as the Earth. This is a major storm that has continued to rage for over 300 years!

Fact Box
Distance from Sun: 778 million kilometers (480 million miles)
Rotation: 9 hrs. 55 min.
Revolution: 11.9 Earth yrs.
Diameter: 143,200 kilometers (89,000 miles)
Gravity: 2 1/2 times greater than Earth's gravity
Orbital Speed: 13 kilometers per second (8 miles per second)
Atmosphere: hydrogen and helium
Temperature: At the top of the cloud cover -130° C (-202° F)
Rings: 3
Satellites: 16
Travel time from Earth:
Jet: 74 years, 3 months
Rocket: 1 year, 9 months
Light Years: 35 minutes
Named for: Roman god, Jupiter (Greek god, Zeus)

1. Jupiter has sixteen satellites. Name the four largest ones discovered by Galileo. (Use an encyclopedia to help find the answer).
   1. *Callisto*
   2. *Ganymede*
   3. *Europa*
   4. *Io*

2. If you could choose between sleeping a full night on Earth or a full night on Jupiter, which one would you pick? Explain your reasoning. (Hint: Think about the rotation of each planet.)
   (Answers will vary, but should contain the fact that Jupiter's night is less than 1/2 of Earth's.)

Saturn

Saturn is composed of gases, like Jupiter, and has no solid surface area. Because of the quick rotation, the planet is flattened somewhat at the poles and wider around the equator. Saturn's atmosphere is different in that it has two wind systems operating in opposite directions. One system blows east to west and the other blows west to east. Saturn has often been considered the most beautiful planet because of its huge ring system. The rings extend almost the same distance as the space between Earth and the Moon! The rings were discovered by Galileo and are composed of rock and ice particles. The particles range in size from tiny pebbles to car-sized boulders. Saturn has more satellites than any other planet.

Fact Box
Distance from Sun: 1.4 billion kilometers (893 million miles)
Rotation: 10 hrs 40 mins
Revolution: 29 1/2 Earth yrs
Diameter: 120,600 kilometers (74,980 miles)
Gravity: 1.07 times Earth's gravity (almost the same)
Orbital Speed: 9.7 kilometers per second (6 miles per second)
Atmosphere: Hydrogen and helium
Temperature: At the top of the cloud cover -176° C (-285° F)
Rings: 8
Satellites: 24 known
Travel time from Earth:
Jet: 150 years, 5 months
Rocket: 3 years, 7 months
Light Years: 1 hr 11 mins
Named for: Roman god, Saturn (Greek god, Cronus)

1. Name the largest moon (satellite) of Saturn. Use the encyclopedia to help find the answer.
   Titan

2. What is unusual about the wind systems of Saturn?
   Two wind systems that operate in opposite directions.

3. In what year did Galileo discover the rings of Saturn? Find the answer in your encyclopedia.
   1610

4. Why is there no solid surface on the planet Saturn?
   The planet is made of gases.

Uranus

Uranus, the seventh planet from the Sun, is also a gas planet. It was discovered in 1781 by Sir William Herschel. Through a telescope, Uranus appears bluish-green in color. A spacecraft, Voyager 2, passed close to this planet in 1986. The pictures that were sent back to Earth showed no geographical features. However, the Voyager 2 did prove that Uranus has eleven rings, ten narrow and one large, composed of particles no smaller than 20 cm (8 in). Uranus has five major satellites that can be spotted through a telescope: Miranda, Ariel, Umbriel, Titania, and Oberon. Space probes have proved that Uranus also has ten much smaller satellites that orbit much closer to the planet. These ten are dark in color and do not reflect as much sunlight as the larger ones.

Fact Box
Distance from Sun: 2.9 billion kilometers (1.8 billion miles)
Rotation: 17.3 hours
Revolution: 84 Earth years
Diameter: 51,100 kilometers (31,750 miles)
Gravity: Almost the same as Earth's (.91 times)
Orbital Speed: 6.8 kilometers per second (4 miles per second)
Atmosphere: 84% hydrogen, 15% helium, 1% other
Temperature: -214 C (-353 F) at top of cloud cover
Rings: 11
Satellites: 15
Travel time from Earth:
Jet: 318 years, 7 months
Rocket: 7 years, 7 months
Light Years: 2 hrs, 30 mins
Named for: Roman god, Uranus

1. Find out which astronomers discovered the five larger moons of Uranus. Also, record the year in which each moon was discovered.

	Discovered by	Year
Miranda	*Gerard Kuiper*	*1948*
Ariel and Umbriel	*William Lassell*	*1851*
Titania and Oberon	*William Herschel*	*1789*

2. Why weren't the ten smaller satellites discovered until a space probe passed close to Uranus?
   They were too small to be seen with our telescopes. They are dark in color closer to the planet, and don't reflect too much light.

© 1996 Kelley Wingate Publications — CD-3728

Answer Key

Neptune

Neptune, the eighth planet from the Sun, is the last gas planet. The planet appears blue because of the methane gas in its atmosphere. Neptune was discovered in 1846 due to the work of two astronomers - John Adams and Jean Leverrier. Several dark spots on the surface are believed to be storms. The largest, called the Great Dark Spot, is the size of Earth. A smaller storm called Scooter has a bright interior that is believed to be wispy clouds. Two satellites, Triton and Nereid, were discovered soon after the planet was identified. Triton is most unusual because it has a retrograde (backwards) orbit around Neptune! The space probe Voyager 2 passed about 5,000 km (3,100 mi) above this planet in 1989, revealing six other smaller satellites.

Fact Box
- Distance from Sun: 4.5 billion kilometers (2.8 billion mi)
- Rotation: 16 hours
- Revolution: 165 Earth years
- Diameter: 49,500 kilometers (30,750 miles)
- Gravity: Almost the same as Earth's gravity (1.15 times)
- Orbital Speed: 5.4 kilometers per second (3 miles per second)
- Atmosphere: hydrogen, helium, and methane
- Temperature: -218 C (-360 F) at cloud cover
- Rings: 5
- Satellites: 8
- Travel time from Earth:
  Jet: 513 years, 2 months
  Rocket: 12 years, 3 months
  Light Years: 4 hrs, 2 mins
- Named for: Roman god, Neptune (Greek god, Poseidon)

1. What is the Great Dark Spot on Neptune?
   A storm the size of the Earth

2. Who discovered Triton and Nereid? In what year?

	Discovered by	Year
Triton	William Lassell	1846
Nereid	Gerard Kuiper	1949

3. What makes Neptune look blue?
   methane gas in the atmosphere.

4. What makes Triton the most unusual satellite in the solar system?
   It orbits backwards.

Pluto

Pluto is the smallest and coolest of the planets. Because it is so far from Earth, there is still much to learn about this planet. It is usually the ninth planet, but not always. Pluto has such an elliptical orbit that, when it is closest to the Sun, it travels inside Neptune's orbit. For about 20 years out of its 248 year revolution, Pluto is actually the eighth planet and Neptune is the ninth! Pluto's surface is a covered with nitrogen, carbon monoxide, and methane frozen into an unusual snow. If you were to step onto Pluto it would feel something like very soft pudding. The surface is red near the equator and bluer at the poles. Pluto's one moon, called Charon, is nearly the same size as the planet it orbits.

Fact Box
- Distance from Sun: 5.9 billion kilometers (3.7 billion mi)
- Rotation: 6 days, 9 hours
- Revolution: 248 Earth years
- Diameter: 2,300 kilometers (1,500 miles)
- Gravity: 1/100 that of Earth's gravity
- Orbital Speed: 4.7 kilometers per second (2.9 mi. per second)
- Atmosphere: nitrogen, carbon monoxide, and methane
- Temperature: -230° C (-382° F)
- Rings: none
- Satellites: 1
- Travel time from Earth:
  Jet: 690 years, 1 months
  Rocket: 16 years, 5 months
  Light Years: 5 hrs, 25 mins.
- Named for: Roman god, Pluto (God of death)

1. Why haven't scientists learned as much about Pluto as they know about other planets?
   It is so far away. No probes have gone there yet

2. What is unusual about the orbit of Pluto?
   It is elliptical. It crosses Neptune's orbit for 20 years out of 248 years.

3. Describe the surface of Pluto in your own words:
   A frozen gas that feels like pudding.
   Answers will vary.

Comparing Planets

Answer the questions below using the information from the Fact Boxes for each planet.

1. Which planet is farther from the Sun: Mars or Venus? **Mars**
   How many kilometers farther? **119 million km.**

2. Which planet has a larger diameter: Neptune or Uranus? **Uranus**
   How many kilometers larger? **1,600 km.**

3. Which planet has a shorter rotation time: Earth or Mars? **Earth**
   How many hours and minutes less? **40 min. less**

4. Which planet is closer to the Sun: Venus or Mercury? **Mercury**
   How many kilometers closer? **50 million km**

5. Which planet has more rings: Saturn or Jupiter? **Saturn**
   How many more? **5 more**

6. Which planet has fewer satellites: Earth or Uranus? **Earth**
   How many fewer? **14 fewer**

7. Which planet has the slower orbital speed: Saturn or Jupiter? **Saturn**
   How many kilometers per second slower? **3 km/sec.**

8. Which planet has the shorter travel time (by jet) from Earth:
   Saturn or Mercury? **Mercury**
   How many years and months less? **139 yrs. 2 months**

9. Which planet has the shorter travel time (in light years) from Earth:
   Jupiter or Venus? **Venus**
   How many minutes shorter? **325 min.**

10. Which planet has the shorter revolution around the Sun:
    Venus or Mercury? **Mercury**
    How many days shorter? **137 days**

Planet Crossword

ACROSS
1. The name of an orbit around the Sun. (REVOLUTION)
4. Caloris Basin is found on this planet. (MERCURY)
6. The Red Planet. (MARS)
7. The Great Red Spot is found on this planet. (JUPITER)
8. The Great Dark Spot and Scooter are found on this planet. (NEPTUNE)
11. The gases that surround a planet are this. (ATMOSPHERE)
16. One complete turn of a planet on its axis (one day). (ROTATION)

DOWN
2. The third planet from the Sun. (EARTH)
3. The sister planet of Earth. (VENUS)
5. The planet with the most satellites. (SATURN)
9. The smallest and coolest planet. (PLUTO)
10. This planet has the moons Miranda, Umbriel, and Ariel. (URANUS)
12. This causes the tides on Earth. (MOON)
13. A heavenly body that orbits a planet. (SATELLITE)
14. Rock and ice particles that create a circle around a planet. (RINGS)
15. The distance around the widest part of a planet. (DIAMETER)

© 1996 Kelley Wingate Publications — CD-3728

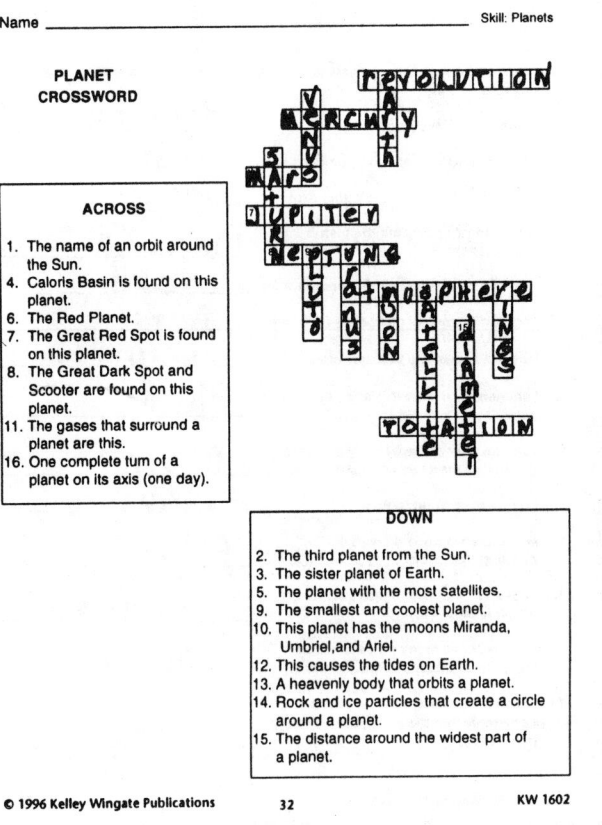

Answer Key

Name _____ **Skill:** Theory

PLANETS
Read the clues below and give the name of what it describes.
Circle the names in the word search. They may go in any direction.

[word search grid]

CLUES
1. Gases that surround a planet. **Atmosphere**
2. Third planet from the Sun. **Earth**
3. A force that attracts matter toward the center of the Earth. **Gravity**
4. The Great Red Spot is found on this planet. **Jupiter**
5. This Red Planet is the subject of many fictional stories. **Mars**
6. Caloris Basin can be found on this planet. **Mercury**
7. The only satellite of the Earth. **Moon**
8. This planet was not discovered until 1846. **Neptune**
9. The smallest and coldest planet of all. **Pluto**
10. To move in an orbit around an object. **Revolution**
11. One complete turn of a planet on its axis. **Rotation**
12. The planet with the most satellites. **Saturn**
13. A bluish-green gas planet discovered in 1781. **Uranus**
14. The sister planet of the Earth. **Venus**

© 1996 Kelley Wingate Publications   33   KW 1602

Name _____ **Skill:** Planets

TRUE OR FALSE?
Read each statement carefully. If the statement is true, put a T on the blank. If the statement is false, put a F on the blank.

1. **F** Mars has three moons: Phobos, Deimos, and Olympus.
2. **F** The Voyager 2 spacecraft passed over Neptune in 1986.
3. **F** Mars is a rocky planet that is much warmer and larger than Earth.
4. **T** Caloris Basin, a large crater area, is found on Mercury.
5. **F** The Great Red Spot is a huge storm on Neptune.
6. **T** Miranda, Ariel, Umbriel, Titania, and Oberon moons of Uranus.
7. **F** Mercury is composed of gases and has no solid surface.
8. **T** Venus appears bright because the clouds reflect so much light.
9. **F** The soil on Mars contains potassium which makes it appear red.
10. **F** The canyon called Valles Marineris is found on Mercury.
11. **T** Neptune was discovered by two astronomers in 1846.
12. **F** Uranus has more rings than any other planet.
13. **T** Jupiter has more satellites than any other planet.
14. **F** Mars is the only planet, other than Earth, that can support life.
15. **T** The tilt of the Earth's axis allows this planet to have seasons.
16. **F** Mercury is the second planet from the Sun.
17. **T** Charon is a moon that is almost as large as the planet it orbits.
18. **T** Pluto is covered with frozen gases that make an unusual snow.

© 1996 Kelley Wingate Publications   34   KW 1602

Name _____ **Skill:** Planets

Who Am I?
Read each clue and decide who or what is the answer. Write your answer on the line. Answers may be used more than once!

1. Triton is one of my many moons. **Neptune**
2. I have 8 rings and at least 24 satellites. **Saturn**
3. I was named for the Greek god, Ares. **Mars**
4. I have no rings and am the planet closest to the Sun. **Mercury**
5. I am 150 million km. away from the Sun. **Earth**
6. The iron in my soil makes me look red. **Mars**
7. I am the first of the four gas planets. **Jupiter**
8. I am usually the ninth planet, but sometimes I am eighth. **Pluto**
9. I am the only celestial body (besides Earth) that man has walked on. **Moon**
10. I rotate from West to East. **Venus**
11. My surface is covered with a pudding-like snow. **Pluto**
12. I have 15 satellites, the largest of which is Miranda. **Uranus**
13. I am flattened at my poles because I rotate so quickly (10 hrs, 40 mins). **Saturn**
14. I have mountains, valleys, plains, and continents like the Earth, but I do not have water. **Venus**

© 1996 Kelley Wingate Publications   35   KW 1602

Name _____ **Skill:** Planets

Rank the Planets
List the planets in order for the information given.

Alphabetical Order	Size (Largest to Smallest)	Temperature (Hottest at any time)
1. Earth	1. Jupiter	1. Venus
2. Jupiter	2. Saturn	2. Mercury
3. Mars	3. Uranus	3. Earth
4. Mercury	4. Neptune	4. Mars
5. Neptune	5. Earth	5. Jupiter
6. Pluto	6. Venus	6. Saturn
7. Saturn	7. Mars	7. Uranus
8. Uranus	8. Mercury	8. Neptune
9. Venus	9. Pluto	9. Pluto

Rotation (Length of Day)	Revolution (Length of Year)	Distance From Sun
1. Jupiter	1. Mercury	1. Mercury
2. Saturn	2. Venus	2. Venus
3. Neptune	3. Earth	3. Earth
4. Uranus	4. Mars	4. Mars
5. Earth	5. Jupiter	5. Jupiter
6. Mars	6. Saturn	6. Saturn
7. Pluto	7. Uranus	7. Uranus
8. Mercury	8. Neptune	8. Neptune
9. Venus	9. Pluto	9. Pluto

© 1996 Kelley Wingate Publications   36   KW 1602

© 1996 Kelley Wingate Publications   CD-3728

Answer Key

LABEL THE MAP
Label the planets and Sun correctly.

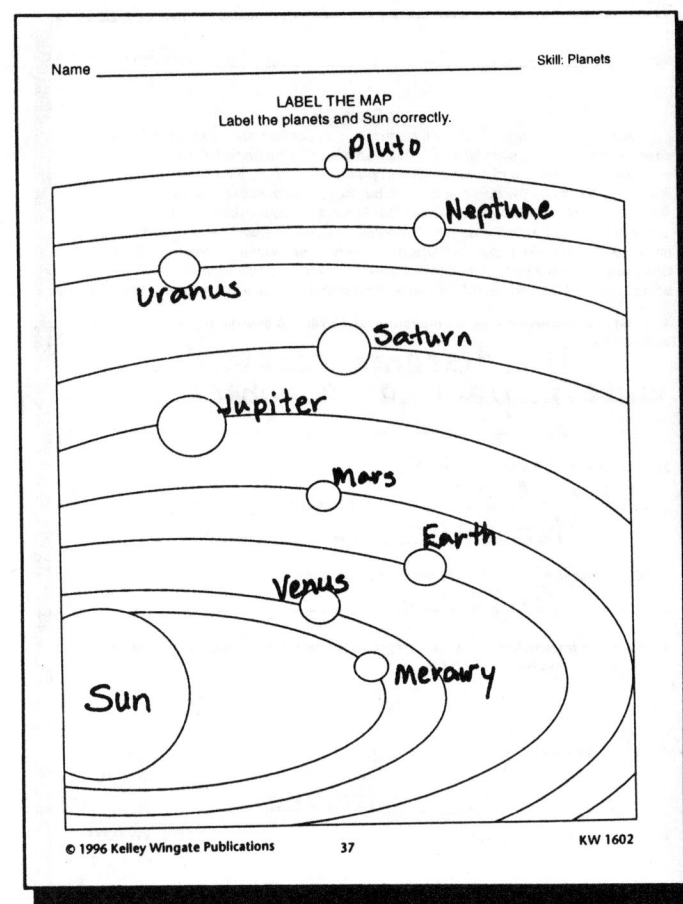

- Pluto
- Neptune
- Uranus
- Saturn
- Jupiter
- Mars
- Earth
- Venus
- Mercury
- Sun

Plan A Vacation
You are planning a vacation on another planet. Name the furthest planet you can visit for each given time and type of transportation.

#	TIME	TRANSPORTATION	PLANET
1.	5 minutes	light years	Mercury
2.	4 years	rocket	Saturn
3.	6 hours	light years	Pluto
4.	8 years	rocket	Uranus
5.	2 months	rocket	Venus
6.	4 months	jet	Moon
7.	9 years	jet	Mars
8.	35 minutes	light years	Jupiter
9.	1 and 1/2 hours	light years	Saturn
10.	3 hours	light years	Uranus
11.	17 years	rocket	Pluto
12.	11 years	jet	Mercury
13.	6 years	jet	Venus
14.	3 months	rocket	Mercury

Matching Planets
Match the planets to the facts about them.

PLANET
- A. Mercury
- B. Venus
- C. Earth
- D. Earth's Moon
- E. Mars
- F. Jupiter
- G. Saturn
- H. Uranus
- I. Neptune
- J. Pluto

FACT
- C — I am a mostly blue planet protected by a layer of clouds.
- H — I have 11 rings and 15 satellites.
- G — My eight rings reach into space about the same distance as it is between the Earth and the Moon.
- I — Triton and Nereid are two of my satellites.
- J — I am named for the Roman god of death.
- E — I have the Valles Marineris, a huge canyon, on my surface.
- J — Charon, my moon, is almost as large as I am.
- C — I am tilted on my axis, which gives me seasons.
- F — I am a gas planet with only three rings.
- B — My rotation is about 243 Earth days.
- D — My dark side is never facing the Earth.
- D — My revolution and rotation are both 27 days.

What is a star?

Somewhere in space a cloud of gas (mainly hydrogen and helium) and dust begins to collect. This cloud is called a nebula (a star in the making). For a few billion years the cloud contracts and grows warmer as more gas and dust are pulled in. When the protostar is dense enough the pressure and heat cause energy releasing reactions (a sort of fiery explosion of atoms) and the cloud begins to burn. The cloud is now a young star. The hydrogen fuses to form helium and energy is radiated as the star expands and burns brightly for another few billion years. As the hydrogen disappears, the star grows older it begins to cool and becomes a red giant (giving off lots of light but not much energy). Smaller stars become white dwarfs and larger stars explode as supernovae (burning brightly and quickly) and finally dies. If the helium core survives the explosion it may become a black hole.

1. What are the stages a star goes through during its lifetime?
   1. A cloud of gases and dust
   2. Protostar
   3. Young star
   4. Red giant
   5. Dwarf or super novae
   6. dies or becomes a black hole

2. What are the two main gases found in stars?
   hydrogen and helium

3. What is a red giant?
   An old star that begins to cool.

4. What two things cause the nebula gases to begin burning?
   Pressure and heat

© 1996 Kelley Wingate Publications — CD-3728

Answer Key

Name _____ Skill: Stars

Star Color

If you have ever watched a fire, you have seen that parts of it are different colors. Scientists know from studying heat that different colors are produced by different degrees of temperature. Stars burning at different temperatures will also produce various colors. The coolest stars are red (also called M-type stars) with surface temperatures of about 3,038 °C (5,500° F). Our Sun is a G-Type star, light orange in color, with a temperature of 5,540° C (10,000° F). Yellow stars have temperatures of about 6,650°C 12,000°F) and are known as F -type. At about 6,900°C (12,500°F) stars turn white and are called A-type stars. The hottest stars are blue, burning between 9,400 and 20,500°C (17,000 and 37,000°F). These blue stars are called O-type.

1. Is it possible for any life to exist on a star? Why or why not?
   No, it is too hot

2. What causes stars to have different colors?
   The temperature.

3. Complete the chart below by filling in the correct information.

STAR CHART

COOLEST				HOTTEST
M-type	G-type	F-type	A-type	O-type
Color: red	Color: orange	Color: yellow	Color: white	Color: blue
Temperature:	Temperature:	Temperature:	Temperature:	Temperature:
C: 3,038	C: 5,540	C: 6,650	C: 6,900	C: 20,500
F: 5,500	F: 10,000	F: 12,000	F: 12,500	F: 37,000

Name _____ Skill: Stars

Star Size

Although most stars are about the same size as our Sun, stars do have different sizes. A few are dwarf stars which are much smaller than the Earth. Others are supergiant stars that are hundreds of times larger than our Sun. It is difficult for scientists to measure the diameter of stars, but they have developed several different ways to determine an approximate size. Our Sun is a medium-sized star with a diameter of about 1,400,000 km (865,000 miles). It would take over 100 Earth's to make the diameter of the Sun. Antares is a supergiant star with a diameter about 330 times larger than our sun! One of the smallest stars known is Van Maanen's Star which is about 9,800 km. (6,100 mi.) wide. That is about the size of the planet Mars.

1. What is a diameter? Use the dictionary and write the answer in your own words.
   The distance across the widest part of a sphere

2. What is the diameter of the Sun?
   1,400,000 kilometers
   865,000 miles

3. Which star is about 330 times larger than the Sun?
   Antares

4. Using the information from the paragraph, correctly label these three stars as Antares, Van Maanen, and Sun:

   o **Van Maanen**   ◯ **Antares**   ◯ **Sun**

Name _____ Skill: Stars

Life Span

Human beings have a life span of about 80 years, but a star usually lives for about ten billion years! The first stage is called a nebula - a cloud of hydrogen and dust. When the hydrogen compacts, changes, and explodes the nebula becomes a baby star. The third stage is a star. During this time the hydrogen continues to fuse into helium and radiates light and heat. At 4.6 billion years old, our Sun is considered a middle-aged star! As the star grows older the hydrogen begins to disappear and the helium grows. At the fourth stage, the helium core cools and shrinks rapidly as the outer shell glows brighter than before but giving off little radiant energy. Now the star becomes a giant or supergiant. During the fifth stage the supergiant cools and contracts until it becomes a white dwarf. This star is slowly dying now that the hydrogen and helium have been nearly used up. Finally, the white dwarf becomes a black dwarf, completely collapsing in on itself.

1. List the six stages in the life of a star:
   1. **Nebula**
   2. **Baby star**
   3. **Star**
   4. **Giant or supergiant**
   5. **White dwarf**
   6. **Black dwarf**

2. What is the life span of most stars?
   10 billion years

3. How old is the Sun and what stage is it in?
   4.6 years
   Star stage

Name _____ Skill: Planets

TRUE OR FALSE?

Read each statement carefully. If the statement is true, put a T on the blank. If the statement is false, put a F on the blank.

T 1. A star is a burning ball of gases and dust.
F 2. The hottest stars are deep red in color.
F 3. A baby star is a cloud of hydrogen and dust.
T 4. Antares is one of the largest known stars.
T 5. A white dwarf is a star very near the end of its life.
F 6. The life span of a star is about three million years.
T 7. Our Sun is a G-type star.
T 8. The blue stars have a temperature up to 20,600° C (37,000° F).
F 9. A nebula is a star that has burned out.
F 10. The van Maanen star is a dwarf.
F 11. Most of the stars are supergiants.
T 12. As hydrogen changes into helium it explodes as fiery energy.
T 13. The core of a giant or supergiant star is mainly helium.
T 14. The last stage of a star's life is black dwarf.
F 15. At the giant stage, a star produces more energy and less light.
T 16. The coolest stars are red in color.
T 17. When a star dies, the helium core may become a black hole.
T 18. Our Sun is about 4.6 billion years old.

Answer Key

Our Sun

The star nearest to the Earth and the center of our solar system is, of course, the Sun. Without its heat and light we could not survive. The Sun rotates on its axis (just like the planets) and averages 25 days for each rotation. The Sun is an average-sized star that is about half way through its ten billion year life. It is yellow in color and is classified as a G-type star. It is composed, or made up of, about 2/3 hydrogen and 1/3 helium. Energy is produced in the core, or middle, of the sun as hydrogen fuses (or alters) because of heat (2 million degrees C or 1 million degrees F) and pressure. The energy rises to the surface of the sun where it cools to about 18,100° C (10,000° F). From the surface the energy is radiated into space as heat and light.

1. Why is the Sun so important to life on Earth?
   It provides heat and light to support life.

2. How long is one solar day (one rotation of the Sun)?
   25 days

3. Where is the Sun's energy produced?
   in the core or middle

4. What is the temperature of the surface of the Sun?
   18,100° Celsius   *10,000°* Fahrenheit

5. What would happen to life on Earth if we were any closer or further away from the sun?
   Closer - We would burn up. Farther - We would not receive enough heat or light for life.

Solar Sunspots

As early as 1612, Galileo noticed several black spots on the Sun. These dark patches, called sunspots, are set back into the surface of the sun and are actually cooler than the surrounding area. Although dark in color, sunspots produce a lot of light. In fact, just one sunspot could light up a dark sky! They may look small compared to the size of the Sun, but sunspots often have a diameter of 18,000 km (11,000 mi.). Sunspots are usually found in groups and it is believed that they have a magnetic attraction toward each other. Scientists have also discovered that there is a cycle to sunspots. They reappear in the same places about every eleven years. Most sunspots form and disappear within a two week period, but some last up to a year.

1. What is a sunspot?
   A dark spot on the Sun.

2. How often does the sunspot cycle repeat itself?
   every 11 years

3. Why do sunspots usually form in groups?
   Scientists think they are attracted by magnetism.

4. What is the average diameter of a sunspot?
   18,000 km   *11,000* mi

5. How long does it normally take for a sunspot to form and disappear?
   about 2 weeks

6. What is the longest period of time a sunspot might last?
   up to one year

Solar Flares

Solar flares are sudden burst of energy in the form of fire that erupts on the Sun's surface. These powerful fire bursts usually occur over an area of a sunspot. No one knows for certain what causes solar flares, but it is believed they are a form of magnetic storm created by solar winds and sunspots. These fire storms can be quite large, covering up to one billion square kilometers (386 million square miles). Usually a flare happens in less than a minute, but it can give off an extreme amount of radiation (waves of energy). This radiation can be harmful to astronauts in space and can even disturb radio waves or compasses on Earth! Scientists track solar flares on the Sun, but they have seen flares on other stars as well.

1. Which are more dangerous to man, sunspots or solar flares? Why?
   Flares give off radiation.

2. What causes solar flares?
   Solar winds and magnetic storms.

3. How are sunspots and solar flares related?
   Most solar flares occur over sunspots.

4. How large can a solar flare become?
   1 billion square km
   386 million square mi

Solar Prominences

The most spectacular event on the Sun is undoubtedly a solar prominence. These fantastic happenings are great rose colored gas clouds that quickly shoot up from the Sun's surface for thousands of kilometers, then loop over and return to the surface. A typical prominence will rise about 40,000 km (25,000 mi) above the surface of the Sun. Solar prominences usually take about an hour from start to finish, although some very large ones have been known to last for several months. Scientists do not fully understand what causes these happenings, but they believe the Sun's magnetic field has something to do with them.

1. The word "solar" means sun and the word "prominent" means immediately noticeable. Why do you think scientists named these shooting fire clouds solar prominences?
   Prominences are so noticeable on the surface of the Sun.

2. In your own words, what is a solar prominence?
   A stream of fire that shoots from the Sun, loops over and returns to the surface of the Sun

3. Why do you think prominences are probably more exciting to see than sunspots or solar flares?
   They are more noticeable, much larger, and do more than spots or flares

4. How long does a typical solar prominence last?
   about 1 hour

© 1996 Kelley Wingate Publications — CD-3728

Answer Key

Solar Eclipse

The Sun is about 400 times larger than the Earth's Moon. It is also about 400 times further from the Earth than the Moon. This strange coincidence makes the Sun and the Moon appear to be about the same size when viewed from the Earth. Every so often the Sun and the Moon are in a direct line with some location on the Earth. Because they look to be the same size, the Moon actually blocks out the Sun's light for a short time. The Earth becomes almost as dark as night as the Moon's shadow (about 160 km. or 100 mi. across) passes over the land. This event is called a total solar eclipse (all of the Sun is blocked). Other parts of the Earth may see the Moon block only part of the Sun as the two orbits cross paths. This is known as a partial solar eclipse (part of the Sun is blocked). If you were to live in the same city forever, you would see a total eclipse only once every 400 years!

1. What makes the Moon and the Sun appear to be the same size?
   The Sun is far away, the moon is closer

2. How wide is the shadow cast by the Moon during a total eclipse?
   160 kilometer
   100 miles

3. What is the difference between a total eclipse and a partial eclipse?
   Total Eclipse *The entire Sun is blocked by the moon.*
   Partial Eclipse *Part of the Sun is blocked by the moon.*

4. How often does a total solar eclipse often in the same location?
   every 400 years

TRUE OR FALSE?

Read each statement carefully. If the statement is true, put a T on the blank. If the statement is false, put a F on the blank.

- **T** 1. Our Sun is the largest star in our solar system.
- **F** 2. Sunspots are bits of sun that fly off into space.
- **T** 3. Stars are made of gases and bits of dust.
- **F** 4. Sunspots have a higher temperature than the area around them.
- **T** 5. Galileo was the first astronomer to see solar prominences.
- **T** 6. Solar flares can affect radio signals and compasses on the Earth.
- **F** 7. The shadow of the Moon during an eclipse is 16 km (10 mi.) wide.
- **F** 8. The Moon is blocked by the Sun during a solar eclipse.
- **T** 9. Solar prominences are loops of fire shot from the Sun's surface.
- **T** 10. Flares can be seen on other planets.
- **T** 11. Sunspots, although dark, produce a great amount of light.
- **T** 12. Sunspots follow a cycle of 11 years.
- **F** 13. Our Sun is very old and has become a supergiant.
- **F** 14. The Sun is one of the smallest stars.
- **T** 15. The Sun's magnetic field is probably the cause of flares.
- **F** 16. Sunspots last for several hundred years.
- **T** 17. Solar prominences spurt about 40,000 km. (25,000 mi.) high.
- **T** 18. The Sun is about 400 times larger than the Earth's Moon.

ACROSTIC

Read the clues and figure out what it is naming. Write the word in the spaces provided. (Clue number one goes first and begins with an A.)

```
    N E B U L A
         S U N S P O T S
      F L A R E S
P R O M I N E N C E
      H E L I U M
         A N T A R E S
S U P E R G I A N T
H Y D R O G E N
            D W A R F
        E C L I P S E
```

1. A cloud of hydrogen, helium, and dust.
2. A group of dark spots on the surface of the Sun.
3. Small bursts of fire near sunspots.
4. Tall loop of fire bursting from the surface of the Sun.
5. A gas at the core of an older star.
6. One of the largest known stars.
7. A star near the end of its life that gives off lots of light but not much heat.
8. The gas that composes most of a baby star.
9. A star near the end of its life that has burned up most of its hydrogen and helium.
10. This happens when the Moon comes directly between the Sun and the Earth, casting a huge shadow on the Earth as it blocking out the Sun.

SOLAR SCRAMBLE

The Sun was so hot yesterday that it scrambled all the letters in the words below. Unscramble the letters and write the "star" words below each sun.

- BLEUNA — *Nebula*
- MELUIH — *Helium*
- GRPUSNEATI — *Supergiant*
- WFADR — *dwarf*
- ERAANST — *Antares*
- TOSSPUN — *sunspot*
- GOHEYRDN — *Hydrogen*
- NIMPCROEEN — *Prominence*
- PLEICSE — *Eclipse*

and
Answer Key

The Milky Way

Our solar system (the Sun and planets) is very large to us, but it is only one small part of our universe. There are several billion other stars in space that we know very little about! When early astronomers looked at the sky they noticed a huge band of stars that looked somewhat like a road of milk. They began to call this band "gala" which is the Greek word for milk. From this term we developed the word "galaxy", which means a collection of stars. The band of stars has become known as the Milky Way, another name for our galaxy. The Milky Way is so large that it must be measured in light years (the distance a ray of light can travel in one year). Light travels at 300,000 km. (186,000 mi.) per second! The Milky Way is about 100,000 light years in diameter. If we fit a model of our galaxy on this paper, the Sun would be smaller than the dot at the end of this sentence and the Earth could only be seen through a microscope.

1. What is a galaxy?
   A collection of stars

2. Why was our galaxy named after milk?
   The stars looked like a white road of milk in the night sky

3. What is a light year?
   The distance a ray of light can travel in one year.

4. Why is the galaxy measured in light years rather than kilometers or miles?
   It is too large to measure in kilometers or miles.

Novas and Supernovas

The Latin word "novus" means new. In the early days of astronomy, a star which suddenly seemed to appear was called a nova (new star). Today we know that a nova is actually an old star that is burning out. The surface of the star suddenly explodes, creating a light that may be thousands of times brighter than the original star. After a few days, months, or years the star returns to its original brightness. This may happen several times before the star becomes a dwarf. There have been hundreds of novas in our galaxy during the past 100 years. Sometimes a star explodes completely, creating a glow that is billions of times brighter than the original star. These explosions are called supernovas, occurring only once every few hundred years. In 1987 a supernova was observed in another galaxy called the Large Magellanic Cloud.

1. Why did early astronomers call a suddenly glowing star a nova?
   It was the first time they saw it so they believed it was a new star.

2. What happens to a nova after it becomes so bright?
   It returns to its original brightness and becomes a dwarf.

3. List two ways in which novas and supernovas are different:
   1. *A supernova explodes completely.*
   2. *A supernova is bigger and brighter.*

4. What happens to a supernova after it becomes so bright?
   It explodes completely

Black Holes

Imagine a star many times larger than our Sun. Its gravity is powerful, pulling in any mass (like gases and dust) that comes too near. The gravity of this star continues to grow until it becomes even stronger than the speed of light. At that point the star would collapse into itself, leaving a large black hole that would capture anything near it. As early as 1798 the theory was developed that these black holes could occur, but it has never been proven. The Hubble Space Telescope photographed a disk of dust and gas in the center of galaxy M87. This disk is believed to be a black hole, but no one knows for sure. Black holes are one of the many mysteries of space yet to be solved.

1. Why do scientists consider black holes to be mysteries?
   They cannot prove that black holes exist.

2. Is a black hole really a hole? Explain.
   No, it is not a hole. It is a densely packed ball of gas, dust, and other matter.

3. In which galaxy have scientist found what they think is a black hole?
   galaxy M87

4. Explain this statement: Black holes are only theories (ideas).
   No one has ever proved that black holes exist, but there is a lot of evidence to suggest that they probably do.

NUMBER MATCHING
Match the numbers on the left with the facts about them on the right.

A. 186,000 miles per second — D temperature of the coolest star
B. 10 billion years — G diameter of the Milky Way Galaxy
C. 4.6 billion years — B life span of a star
D. 3,038° C (5,500° F) — N Galileo discovered sunspots
E. 20,500° C (37,000° F) — Q height of a solar prominence
F. 1987 — O rotation and revolution of the Moon
G. 100,000 light years — J size of Antares
H. 2,000,000° C (1,100,000° F) — F supernova discovered in another galaxy
I. 1798 — H temperature at the core of the Sun
J. 330 times larger than the Sun — I theory of black holes was developed
K. 11 years — M size of the Moon
L. 2/3 hydrogen, 1/3 helium — A speed of light
M. 1/4 the size of the Earth — K cycle for sunspots
N. 1612 — L amount of gases in the Sun
O. 27 days — E temperature of the hottest stars
P. 160 km (100 mi) — C age of our Sun
Q. 40,000 km (25,000 miles) above the Sun — P diameter of Moon's shadow during total eclipse

---

© 1996 Kelley Wingate Publications
CD-3728

Answer Key

TRUE OR FALSE?

Read each statement carefully. If the statement is true, put a T on the blank. If the statement is false, put a F on the blank.

T 1. The Milky Way Galaxy is 100,000 light years in diameter.
F 2. The Sun and its planets are one galaxy.
T 3. Each galaxy has millions of stars.
T 4. "Gala" is the Greek word for milk.
T 5. Our solar system is part of the Milky Way Galaxy.
F 6. "Novus" or nova means old.
F 7. A nova is a star that is just beginning its life.
T 8. Scientists have seen hundreds of novas in the last 100 years.
T 9. A nova may shine brightly then fade several times before it dies.
T 10. Supernovas are stars that explode completely.
T 11. In 1987, scientists found a supernova in galaxy M87.
F 12. There are at least three black holes that scientists have studied.
T 13. Black holes are stars that absorb their own light.
F 14. Scientists discovered a black hole in 1798.
T 15. The Hubble Telescope photographed what could be a black hole.
F 16. Black holes are only unproven theories.
F 17. The gravity of a black hole is not as strong as the speed of light.
F 18. A black hole is a large hole.

STAR CROSSWORD

Crossword answers:
- HYDROGEN
- HELIUM
- MTYPE
- PROMINENCE
- ECLIPSE
- FLARE
- SUPERNOVA
- ANTARES
- GALAXY
- BLACKHOLE
- MILKYWAY
- NEBULA
- SUN

ACROSS
2. A gas that makes up the majority of a new star.
3. These stars are red and the coolest.
4. When the Moon has moved directly between the Sun and the Earth.
7. An exploding star is called this.
8. The name of the largest known star.
11. What we call a vast group of stars.
13. A star that has collapsed on itself.
14. The name of our galaxy.

DOWN
1. Loops of fire that burst from the Sun.
2. A gas that makes up the majority of an older star.
5. The fourth stage of a star (when it glows brightly).
6. Bursts of solar energy that occur over or near a sunspot.
9. A cloud of hydrogen, dust, and helium.
10. The name of the star in our solar system.
12. An old star that flares brightly as it begins to die.

STARS

Read the clues below and give the name of what it describes. Circle the names in the word search. They may go in any direction.

CLUES
1. The largest known star. **Antares**
2. An older star that has become small. **dwarf**
3. This happens when the Sun, Moon, and Earth are in a direct line. **eclipse**
4. This gives off radiation that disturbs radio waves on Earth. **Flare**
5. What the Milky Way is. **galaxy**
6. The main gas in older stars. **helium**
7. The main gas in new stars. **hydrogen**
8. The 1st stage of a star. **nebula**
9. An old star that suddenly explodes. **nova**
10. Hot blue stars. **O-type**
11. The most spectacular event on the Sun. **prominence**
12. A middle aged G-type star in our solar system. **sun**
13. These reappear in the same place on the Sun every eleven years. **sunspot**
14. The average number of years a star exists. **ten billion**

Constellations

Early Greeks noticed that the stars seem to form patterns that make it easy to describe directions or areas of the sky. With a little imagination, groups of stars became pictures. We call these "pictures" constellations. Ptolemy named 48 different groups of stars (mainly characters from Greek mythology). In time, other astronomers added forty more so now every star is part of a constellation. The sky is divided into sections called northern, equatorial (over the equator), and southern. The constellations give astronomers a kind of map of the stars, making it easier to locate any one of of them!

1. What is a constellation?
   A group of stars that make a picture

2. How do constellations make it easier to locate stars?
   It is easier to find a star within a picture than by itself

3. Name the three sections the sky is divided into:
   1. **northern**
   2. **equatorial**
   3. **southern**

4. How many constellations are there? **88** constellations

© 1996 Kelley Wingate Publications — CD-3728

Answer Key

URSA MAJOR

One constellation that is easy to locate is Ursa Major, the Great Bear. It is commonly called the Big Dipper and can be seen in the northern skies all year round. Ursa Major is a group of seven very bright stars that form an almost rectangular shape with a bent handle of three stars. The first two stars in the bowl, Dubhe and Merak, are called Pointers. If you draw an imaginary line from the lower star (Merak) through the upper star and continue until you come to a very bright star, you have located Polaris (the Pole Star or North Star). Polaris remains at the same place in the sky all the time and is very bright. Ancient sailors, among others, used this star as a guide so they always knew the direction in which they were headed.

1. What are two other names for Ursa Major?
   1. Great Bear
   2. Big Dipper

2. Why are Dubhe and Merak called Pointers?
   If you draw a line through these two stars they point to Polaris, the North Star.

3. What are two other names for Polaris?
   1. North Star
   2. Pole Star

4. Why is Polaris an important star?
   It is the only star that stays in the same place in our sky all year

Comets

A comet is a celestial body that orbits the Sun, passing quickly amongst planets on its journey. It appear as a flat ball, sometimes dragging a tail behind it. A comet is made of several gases, water, and dust that are frozen into a kind of dirty snowball. The average diameter of a comet is 100,000 km. (62,000 mi.). When the comet is far from the Sun it travels at about 3,300 km. (2,000 mi.) per hour. As it nears the Sun the speed may increase to over 161,000 km. (100,000 mi.) per hour. About ten new comets are discovered every year. You can usually see a comet without using a telescope about once every three years.

1. What is a comet made of?
   Frozen gas, water, and dust

2. Compare a planet and a comet. How are they alike? Different?
   Alike: Both are round and both orbit the Sun
   Different: A comet is made of gas, water, and dust. A comet has a tail.

3. Does a comet always travel at the same speed? Explain.
   No, it goes faster as it nears the Sun.

4. Pretend that you are the first person to discover a new comet. What would you name it and why?
   (Answers will vary).

Parts of a Comet

A comet is much like a dirty ice ball. The center, or nucleus, of this ball is composed of frozen gases, water, and dirt. As it travels through space, the outside of the nucleus warms and gives off a cloud of gas that surrounds it. This cloud is called the comet and averages 100,000 km. (62,000 mi.) in diameter. As the comet nears the Sun it travels even faster and the coma becomes much larger. The solar winds push against the coma, creating a tail that streams from the nucleus. As the comet nears the Sun, the tail is behind it. However, as the comet is moving away from the Sun, the tail is in front! Scientists believe the strong solar winds are responsible for this fact.

1. Name the three parts of a comet.
   1. nucleus
   2. coma
   3. tail

2. Label the three parts of the comet:
   1. coma
   2. nucleus
   3. tail

3. Why does the tail always point away from the Sun?
   Solar winds push it away

4. How long is a comet's tail?
   100,000 km   62,000 mi.

Halley's Comet

In 1682, Sir Edmund Halley saw a comet. He developed the theory that comets orbit the Sun, as planets do, and that this comet was the same one sighted in 1531 and 1607. Halley predicted that the comet would return in 1758. It did, and the comet was named Halley's Comet in honor of Sir Edmund. Halley's comet appears in our sky every 76 years. The comet makes an orbit of 7 billion miles that takes it just past Neptune. The tail of Halley's Comet has been measured at about 150 million km. (93 million mi.) which is the same distance as the Earth to the Sun!

1. Why is this comet called Halley's Comet?
   Halley figured out the pull of comets

2. How often is this comet seen by Earth?
   every 76 years

3. How many times has Halley's Comet been visible to Earth since the year 1758? (Hint: How often does it appear?)
   3 times

4. How long is the tail of this comet?
   150 million kilometers   93 million miles

5. In what year will Halley's Comet return to the Earth's sky?
   2062

Answer Key

Asteroids

Asteroids, sometimes called minor planets, are small solid objects that also orbit the Sun. They are made of iron, nickel, stone, or any combination of these. There are thousands of asteroids and over 5,000 of them are found between Mars and Jupiter in what is called the Asteroid Belt. Asteroids range in size from many kilometers to a few meters. The largest known asteroid is called Ceres and measures 785 km. (500 mi.) in diameter. It is believed that asteroids are actually parts of the solar system that never joined together as planets. Most of the ones known today are probably pieces of larger ones that smashed against each other and broke apart. Scientists have found evidence of the Earth being hit by asteroids about the time of the ice age (2.3 million years ago). In more recent years asteroids have passed as close as 770,000 km. (478,000 mi.), but there have been no collisions.

1. Why are asteroids called minor planets?
   They are parts of a solar system that never joined as planets but are fairly large.

2. Where can we find the Asteroid Belt?
   Between Mars and Jupiter

3. What are asteroids?
   Pieces of iron, nickel, stone or a combination that orbits the Sun.

4. Look in the encyclopedia and find the names of two other large asteroids.
   1. *Palles   Gaspra*
   2. *Vesta*

Meteoroids, Meteors, and Meteorites

Meteoroids are small bits of rock and dust that are found in space. They are often pulled into the Earth's atmosphere by the force of gravity. Once a meteoroid enters the atmosphere it begins to burn, creating a streak of light in the sky. At this point it is called a meteor, although it is commonly called a falling or shooting star. Although many meteors burn completely before they reach the ground, not all of them do. If a meteor reaches the ground it is called a meteorite. About 500 large meteorites make it to the ground each year, but only about 5 of those are ever recovered.

1. Describe the differences among these three words:

   meteoroid *Rock and dust in space*

   meteor *Burning rock or dust in Earth's atmosphere*

   meteorite *Meteor that reaches the ground*

2. Why are meteors sometimes called falling stars?
   They look like stars falling through the sky.

3. What causes a meteoroid to enter the Earth's atmosphere?
   Gravity pulls them in.

4. Meteorites have caused craters on the Moon. Why do you think the Earth does not have as many as the Moon?
   The Earth's atmosphere burns up most of them.

Meteor Showers

Almost any night, especially after midnight, it is possible to see a couple meteors as they fall to Earth. Sometimes so many fall at once that it is called a meteor shower. This happens when the Earth passes through a wave of meteoroids, believed to be left by a passing comet. The average meteor in such a shower is about the size of a grain of sand. Meteor showers are named after the constellation from which they seem to fall. The most brilliant shower is named Leonid because it seems to come from the constellation Leo, the lion. In 1966 over 1000 meteors per second were seen in the Leonid meteor shower!

1. What causes a meteor shower?
   The Earth passes through a wave of meteoroids.

2. How are meteor showers named?
   They are named for the constellation they seem to fall from.

3. What is the average size of a meteor in a shower?
   About as large as a grain of sand.

4. What was so special about the Leonid Meteor Shower of 1966?
   Over 1,000 meteors per second were seen.

5. Where do the bands of meteoroids come from?
   They are probably left by a passing comet.

TRUE OR FALSE?

Read each statement carefully. If the statement is true, put a T on the blank. If the statement is false, put a F on the blank.

- **F** 1. Ptolemy named all of the 88 constellations.
- **F** 2. Ursa Major is the largest asteroid ever discovered.
- **F** 3. Polaris is part of the Big Dipper constellation.
- **T** 4. The North Star does not change position in our sky.
- **F** 5. Dubhe and Merak are constellations.
- **T** 6. A comet increases in speed as it nears the Sun.
- **F** 7. A comet tail always points toward the Sun.
- **T** 8. Comets are made of gases, water, and dust.
- **T** 9. About 100 new comets are discovered each year.
- **T** 10. Nucleus, coma, and tail are parts of a comet.
- **F** 11. The heat from the Sun pushes the comet tail backward.
- **F** 12. Halley's comet appears every 96 years.
- **T** 13. Asteroids are made of nickel, iron, or stone.
- **F** 14. The Asteroid Belt is found between Jupiter and Saturn.
- **F** 15. Meteoroids are meteors that hit the ground.
- **T** 16. Meteoroids are probably left in space by passing comets.
- **F** 17. Meteoroids are also called shooting stars.
- **F** 18. Less than five meteorites make it to the ground each year.

© 1996 Kelley Wingate Publications

Answer Key

Who Am I? (p. 69)
Read each clue and decide who or what is the answer. Write your answer on the line. Answers may be used more than once!

1. I named 48 of the constellations. — **Ptolemy**
2. I can be seen every 76 years. — **Halley's Comet**
3. I am sometimes called a minor planet. — **asteroid**
4. I am a small particle that burns as I fall to Earth. — **meteor**
5. I am a group of stars that form a picture. — **constellation**
6. I am made of frozen ice, gas, and dust. — **comet**
7. I am the largest known asteroid. — **Ceres**
8. I am small pieces of rock and dust left in space by a comet. — **meteoroid**
9. In 1682 I saw a fuzzy star with a tail and predicted when it would be seen again. — **Edmund Halley**
10. I am the core of ice and dust in a comet. — **nucleus**
11. I was once a meteoroid and will soon be a meteorite. — **meteor**
12. I am what causes a comet's tail to point away from the Sun. — **solar wind**
13. The speed of my orbit increases as I come close to the Sun. — **comet**
14. There are 88 of us and most of us were named for characters in Greek mythology. — **constellations**

Acrostic (p. 70)
Read the clues and figure out the name of the astronomer. Write his name in the spaces provided. (Clue number one goes first, two second, and so on.)

1. A**S**TEROID
2. M**E**TEORITE
3. P**O**LARIS
4. S**O**LAR WINDS
5. CON**S**TELLATION
6. COM**E**T
7. URSA M**A**JOR
8. CE**R**ES
9. NU**C**LEUS
10. **H**ALLEY

1. Iron, nickel, or stone pieces that orbit the Sun.
2. A small bit of burned rock or dust that falls to the surface of the Earth.
3. The North Star used by early sailors to figure out directions at night.
4. A force that pushes the tail of a comet away from the Sun.
5. A group of stars that form a picture.
6. A ball of ice that orbits the Sun.
7. Another name for the Big Dipper or Great Bear.
8. The name of the largest known asteroid.
9. The frozen center of a comet that is surrounded by the coma.
10. The man who predicted a 76 year cycle for a particular comet.

Celestial Sights (p. 71)
Read the clues below and give the name of what it describes. Circle the names in the word search.

1. Solid objects that orbit the Sun in a beltway between Jupiter and Mars. — **asteroids**
2. The largest known asteroid. — **Ceres**
3. The cloud that surrounds a comet. — **coma**
4. A ball of frozen gas, water, and dust. — **comet**
5. A group of stars that form a picture. — **constellation**
6. He predicted when a comet would return. — **Halley**
7. "Falling stars". — **meteor**
8. A meteor that hits the surface of the Earth. — **meteorite**
9. Small bits of rock or dust found in space. — **meteoroid**
10. The North Star. — **Polaris**
11. The man who named many of the constellations. — **Ptolemy**
12. The force that pushes a comet tail away from the Sun. — **solar wind**
13. The part of a comet that streams behind the head. — **tail**
14. A constellation that looks like a water dipper. — **Ursa major**

History of Rockets (p. 72)

Throughout time man has dreamed of space travel. Many cultures have folk tales and stories that tell of gods coming from the stars. Science fiction books are full of tales about travel to other planets and star systems. The rocket seemed the most likely way to turn these dreams into reality. China was known to have rockets as early as AD 1232. These "rockets" were arrows propelled by gunpowder. In 1883 Konstantin Tsiolkovsky, a Russian schoolteacher, wrote a detailed theory about the travel of liquid propelled rockets through space. In 1926, Robert Goddard successfully launched the first rocket. It rose only 41 feet into the air, but proved that rocket travel was possible. During the 1930's, Germans and Russians began to experiment as well. The Russians launched the first rocket to leave the Earth's atmosphere in 1957.

1. Which country was the first to produce rockets? — **China**
2. Name the Russian teacher that first developed the idea of using rockets in space. — **Konstantin Tsiolkovsky**
3. What made Goddard's 1926 launching a success even though the rocket barely left the ground? — **This proved that rocket travel was possible**
4. Which country was the first to launch a rocket into space? In what year? — **Russia in 1957**
5. Look in the encyclopedia to find the name of the first rocket in space. — **Sputnik**

© 1996 Kelley Wingate Publications — CD-3728

Answer Key

Space Race

The Space Age began on October 4, 1957 when the Soviet Union launched their rocket, Sputnik, that orbited the Earth for 92 days. Three months later the United States sent Explorer 1 to orbit the Earth. The race for space exploration was on! In October 1959 the Soviet Moon probe, Luna 3, sent back the first pictures of the dark side of the Moon. In 1961, Soviet cosmonaut Yuri Gagarin was the first man to leave the Earth's atmosphere and orbit our planet. An unmanned landing on the moon was accomplished in 1966 with the Soviet craft Luna 9. The US Apollo 8 carried men around the moon in 1968 and Apollo 11 (1969) carried the first men to walk on the surface of the Moon. Manned flights have not gone further than the Moon, but many unmanned probes have been sent close to planets like Saturn, Uranus, and Neptune.

1. Which two countries were the leaders in early space exploration?
   Russia and United States

2. Write the name of the country that was first in each event listed below:
   - first to put a man in space: **Russia**
   - first to walk on the Moon: **United States**
   - first to orbit the Moon: **Russia**
   - first to orbit the Earth: **Russia**
   - first to land on the Moon: **Russia**

3. What was the name of the first man in space?
   Yuri Gagarin

4. Where have unmanned space probes been sent?
   Saturn, Uranus, Neptune

5. What is the name of the first man to walk on the Moon? (Use an encyclopedia to find the answer.)
   Neil Armstrong

First Astronauts

The first astronauts and cosmonauts (Russian astronauts) were neither men nor women. They were animals! Scientists were not sure what affect space travel would have on humans. In 1948 the United States began to send monkeys up in rockets to see what happened as they left and returned to Earth's atmosphere. The Soviet craft, Sputnik 2, was launched in November 1957 with a dog named Laika aboard. For the next several years dogs and monkeys were the first "astronauts" in space. In 1961, Yuri Gagarin of the Soviet Union became the first man to go outside the Earth's atmosphere. Most of the first astronauts were men selected from the military. The first women were selected for the space shuttle programs, although Russia used women much earlier in their programs.

1. Why were animals the first living beings launched in space?
   Scientists weren't sure how space travel would affect humans so animals were used.

2. What was the first animal to orbit the Earth? What was her name?
   A dog named Laika.

3. What is a cosmonaut?
   A Russian astronaut

4. When did U.S. women first become astronauts?
   When the space shuttle program began

5. How many years were animals used before the first human went into space?
   13 years

Early Astronauts

Yuri Gagarin was the first human being to orbit the Earth. He was aboard Sputnik 2, a Soviet spacecraft. Alan Shepard became the first United States astronaut in 1961. He travelled in space only fifteen minutes before he splashed down in the ocean. John Glenn (U.S.) orbited the Earth in 1962. Scientists watched him carefully as he ate a meal to see how food was affected in space. Valentina Tereshkova, a Soviet cosmonaut, became the first woman in space when she flew Vostok 6 around the Earth 48 times in June, 1963. In 1969, Neil Armstrong (U.S.) became the first man to walk on the surface of the Moon. As his foot touched the Moon he said, "That's one small step for man, one giant leap for mankind."

1. Who was the first woman to travel in space?
   Valentina Tereshkova

2. What is the name of the first person to walk on the Moon? When did he do this?
   Neil Armstrong in 1969

3. Which astronaut was tested eating in space?
   John Glenn

4. Why do you think the first astronaut in space only orbited the Earth instead of heading directly to the Moon?
   (Answers will vary.) Unsafe, couldn't reach the moon yet, etc.

5. What do you think Neil Armstrong meant by his statement?
   (Answers will vary.) Importance of the event was for all men.

Space Time Line

Choose each event from the box at the bottom of this page and write it over the correct date on this timeline.

1232	1926	1948
rocket arrows	Goddard's rocket	monkeys in space

1957 October	1957 November	1959
Sputnik	Laika	Photos: Back of moon

1961	1962	1963
Yuri Gagarin	John Glenn eats in space	Valentina Tereshkova

1966	1968	1969
Luna 9	Apollo 8	Man walks on moon

Valentina Tereshkova · Sputnik · Photos: back of Moon
rocket arrows · Laika · Goddard's rocket
Man walks on Moon · Yuri Gagarin · monkeys in space
Luna 9 · John Glenn eats in space · Apollo 8

© 1996 Kelley Wingate Publications — CD-3728

Answer Key

Who Am I?
Match the names to the facts about them.

NAMES
- M — Neil Armstrong
- E — Sputnik
- A — Rocket arrows
- G — monkeys
- D — Soviet Union
- C — Robert Goddard
- L — Valentina Tereshkova
- F — Apollo 11
- I — Yuri Gagarin
- N — Luna 3
- J — Alan Shepard
- B — Konstantin Tsiolkovsky
- K — John Glenn
- H — Laika

FACTS
A. The Chinese used gunpowder to shoot me at enemies.
B. I wrote about liquid propelled rockets in 1883.
C. I launched the first rocket 41 feet in the air.
D. We put the first man in space.
E. I was the first Soviet rocket in space.
F. I carried the first astronauts to walk on the Moon.
G. We were the first animals to travel in rockets.
H. I was the first dog to orbit the Earth.
I. I was the first man in space.
J. I was the first United States astronaut in space.
K. I was tested eating in space.
L. I was the first woman cosmonaut to travel in space.
M. I was the first person to walk on the Moon.
N. I took the first pictures of the back of the Moon.

The Space Shuttle

1. Name the four parts of the Space Shuttle.
   1. Orbiter
   2. Rocket boosters
   3. External fuel tank
   4. Three main engines

2. Which part of the Space Shuttle cannot be reused? Why not?
   External fuel tank because it burns as it enters the Earth's atmosphere.

3. What protects the Space Shuttle as it re-enters the Earth's atmosphere?
   Insulated silicon tiles

The Crew Cabin

1. How large is the crew cabin?
   71.5 cubic meters

2. Why don't astronauts need to wear space suits while in the Shuttle?
   The air is pressurized.

3. What is on the top floor of the crew cabin?
   The flight deck where the commander and pilot sit.

4. Where does an astronaut sleep during a mission?
   In a bunk or sleeping bag on the bottom level.

5. Name the three kinds of people aboard a Shuttle.
   1. commander/pilot
   2. mission specialists
   3. crew members

Spacelab

1. What is Spacelab?
   An international space project to conduct experiments.

2. Where is Spacelab carried on the Space Shuttle?
   In the cargo bay

3. Who are payload specialists?
   Scientists who are equipment experts.

4. Name four of the countries that helped develop Spacelab. (Use the encyclopedia to find the answer).
   1. (Belgium, Denmark, France,
   2. Germany, Italy, Spain,
   3. the Netherlands, Switzerland,
   4. United Kingdom, Austria)

5. What kind of experiment would you like to conduct in the Spacelab?
   (Answers will vary.)

© 1996 Kelley Wingate Publications — CD-3728

Answer Key

Space Stations

A space station is like a city in a satellite that orbits the Earth. It is designed for astronauts spending long periods of time in space. It contains a laboratory, workshop, and living quarters. Other spacecraft can come and go from a space station, so it acts as a sort of space age "hotel". The first space station was Salyut 1 launched by the Soviets in 1971. That year, three cosmonauts spent 23 days there. Since then the Soviets built another space station called Mir, where a cosmonaut set a record for staying 439 days in space! A robot spacecraft delivers any needed supplies and mail.

1. What is the difference between a space station and the Space Shuttle?
   The space station is a satellite like a city in space. The Space shuttle transports goods to and from the Earth.

2. What is the longest time ever spent in space?
   439 days (Space station Mir)

3. Why do you think space stations are important in the development of space travel?
   (Answers will vary.)

4. If you were to live on a space station for the next year, what one item from home would you like to take? Why?
   (Answers will vary.)

Lift Off!

How can something as large as a Space Shuttle get enough power to push itself into space? Well, lots of fuel is a good guess! The three main engines of the orbiter fire first, building up power. The two solid rocket boosters (on the side of the external fuel tank) ignite and lift off takes place. After about two minutes the solid rockets blaze out. They fall away from the tank and parachutes carry them gently to the ocean. They are recovered and used again on another flight. Eight more minutes pass before the external tank of fuel is burned completely. This tank is also dropped, but it breaks into thousands of pieces before it plunges into the ocean. Finally, two of the main engines push the orbiter into a path around the Earth.

1. Which part of the Shuttle is recovered from the ocean to be reused?
   The solid rocket boosters

2. Which part of the Shuttle is never recovered?
   The external fuel tank

3. How many fuel tanks are necessary to lift off a Space Shuttle? Name them.
   Three main engines, Solid rocket boosters, External fuel tank

4. How is a Space Shuttle more economical to fly than earlier rockets?
   It can be reused. Earlier rockets could not be reused.

Microgravity

One of the fun yet frustrating things about traveling in space is a lack of gravity. Although there is some gravity present, the body feels weightless and things seem to "float" in the air. Scientists call this microgravity. Microgravity creates some changes in the human body. Blood shifts from the lower to the upper part of the body. This causes eyes to look smaller and forehead wrinkles to disappear while the waist and feet become smaller. In microgravity a person is taller than they are on Earth because gravity is no longer pushing down the discs in the spine! Microgravity presents problems, too. Astronauts usually get the sniffles, become dizzy, and are often nauseated. The heart, bones, and muscles do not have to work as hard and can become weakened after long periods in microgravity. These problems disappear after a few days back on Earth.

1. What is microgravity?
   A feeling of weightlessness

2. What happens to the blood in your body during a space flight?
   It shifts to the upper part of your body.

3. Why is a person taller in space than on Earth?
   Gravity no longer pushes down the discs in the spine.

4. Which side effects of microgravity are most harmful to your health over a long period of time?
   Heart, bones, and muscles weaken because they don't work as hard in space as they do on Earth.

Eating in Space

A Space Shuttle contains a galley, or kitchen, for food preparation. The pantry, oven, trays, and hot and cold water are kept in the galley. Astronauts eat three meals each day, selecting from over 100 foods and twenty beverages kept on board! Menus are made and repeated every seven days. There is no refrigerator on board so food must be dehydrated (water is removed and must be replaced before eating), thermostabilized (heated then put in foil pouches), irradiated (preserved by radiation), dried, or natural. Each crew member takes a turn at galley duty. A typical meal takes about 30 minutes to prepare.

1. What is a galley?
   a kitchen

2. How many meals does one astronaut eat in one week?
   21 meals

3. Why wouldn't you find fresh milk, butter, or ice in the Shuttle galley?
   There is no refrigeration.

4. Name five methods of food preparation for travel in space:
   1. *Dehydrated*
   2. *Thermostabilized*
   3. *Irradiated*
   4. *Dried*
   5. *Natural*

Answer Key

Name _____ Skill: Space Exploration

Dinner Time!

Eating in space is not exactly easy. The first problem is staying at the table! Astronauts must strap themselves down to keep from floating around the cabin. Food cannot be served on plates or in bowls as it is on Earth. Food must be served in containers that keep it from floating about the cabin. Getting a drink is another problem. It is not possible to pour water or juice into a glass. Microgravity causes liquids to break up into balls that float around until they hit something and spread out. Astronauts must be very careful not to let crumbs or liquids "spill" because they may get into equipment and cause serious problems. They must eat slowly and carefully to avoid spills.

1. Why must astronauts be so careful about spills and crumbs?

   Spills and crumbs may get into equipment and cause serious problems.

2. How do liquids react when poured in microgravity?

   They break into small balls and float around.

3. List three foods you had today that would not be good for space:
   1. _____
   2. (Answers will vary.)
   3. *Any refrigerated foods.*

Name _____ Skill: Space Exploration

Space Age Food

Below are listed some foods commonly found on a Space Shuttle. Rename the foods using space age words from our solar system. For example, instead of ice cream you could have "Polar Snow".

Common Name	Space Age Name
bread	(Answers will vary)
dried apricots	
scrambled eggs	
pecan cookies	
hot dogs	
macaroni and cheese	
tuna fish	
pudding	
orange drink	
beef and gravy	
applesauce	
jam or jelly	
bananas	

Name _____ Skill: Space Exploration

Clothing in Space

The Space Shuttle carries two spacesuits designed for any work outside of the pressurized cabin areas. These suits, called EMU (extravehicular mobility unit), allow astronauts to work in the cargo bay, outside the Shuttle, or on the Moon. The EMU has three parts: the liner, the pressure vessel, and the primary life-support system backpack. The liner is like long underwear filled with tubes of water to keep the person cool. The pressure vessel includes the outer jacket, pants, gloves, and helmet. These pieces come in different sizes that can all be attached to make the suit. The primary life-support system backpack (PLSS) is worn on the back and connects to the suit to supply oxygen. The PLSS holds enough oxygen for seven hours of work outside the cabin.

1. When does an astronaut wear an EMU?

   To work in the cargo bay, work outside the shuttle, or walk on the moon.

2. What do the letters EMU and PLSS mean?

   EMU *Extravehicular Mobility Unit*
   PLSS *Primary Life-Support System*

3. What is unusual about the liner or "long johns" of an EMU?

   They have tubes of water in the liner to keep the person cool.

4. How do astronauts make the EMU's fit the different size bodies?

   The pieces come in different sizes that can be attached to each other.

5. What does the PLSS carry?

   an oxygen supply

Name _____ Skill: Space Exploration

Keeping Space Clean

On Earth it is easy to take out the trash, but how is that done in space? Trash and garbage quickly breed unhealthy germs that can rapidly spread in a closed area like the Space Shuttle. The cabin must be kept clean to avoid this problem. Food trays are wiped clean with a wet paper cloth. Empty food containers, leftovers, and paper wipes are put into plastic bags that are then tightly sealed. The sealed bags are tossed into a special storage locker that will be emptied when the Shuttle returns to Earth. Astronauts must also keep themselves clean. They put on fresh underwear daily, a clean shirt every three days, and clean pants once a week. The soiled laundry is sealed in plastic bags and stored for return to Earth.

1. Why must trash be stored so carefully on the Space Shuttle?

   to avoid spreading unhealthy germs

2. What do astronauts do with left over food and empty containers?

   they are sealed inside plastic bags

3. Where is garbage placed on board the Space Shuttle?

   a special storage locker

4. Why do you think waste materials are brought back to Earth?

   so as not to pollute space
   (Answers may vary.)

5. How could waste and laundry become a problem on a space station?

   It may be hard to find storage for so many waste items.

© 1996 Kelley Wingate Publications — CD-3728

Answer Key

Re-entry

The time has come for the Space Shuttle to return to Earth. Everything that is not needed must go into storage. Cargo bay doors are shut. Seats come out of storage and are bolted to the floor. Cargo bay doors are shut. The crew members put on antigravity suits to help their bodies adjust to the changes during re-entry, then strap themselves in their seats. Engines are used to slow the spacecraft down. The tiles that line the outside of the Shuttle glow red with friction heat as the craft re-enters the Earth's atmosphere. The shuttle slows from 25 times the speed of sound to subsonic speeds within a few minutes. The craft now flies like an airplane as it approaches a runway. After landing, the crew is carefully checked by physicians before they go home.

1. Why must crew members wear special suits for re-entry?
   to help the body adjust to changes during re-entry.

2. What causes the tiles to turn red?
   the friction which causes heat.

3. Below are listed some events of re-entry. Put a one beside the event that happens first, a two for second, and so on.
   - **4** Crew members are strapped into their seats.
   - **7** The Shuttle is flown like a plane.
   - **5** Engines slow down the craft.
   - **2** Seats are bolted to the floor.
   - **6** The Shuttle re-enters the Earth's atmosphere.
   - **1** Things are put into storage.
   - **3** Antigravity suits are put on.

Astronaut Acronyms

An acronym is a word or term made up of the first letters of many words. For example, NASA is an acronym for National Aeronautics and Space Administration. Below is a list of common acronyms used by astronauts. Match the acronym with the full term.

- **H** TDRSS — A. Portable Life-Support System (backpack with oxygen)
- **G** ATU — B. Extravehicular Mobility Unit (Spacesuit with life support)
- **L** IVA — C. Extravehicular Activity (work outside the crew cabin)
- **D** SOMS — D. Shuttle Orbiter Medical System (medical kit with monitors)
- **B** EMU — E. Space Transportation System (shuttle, tank, and boosters)
- **O** MCC — F. High-Order Assembly Language for Shuttle (computers)
- **I** AFB — G. Audio Terminal Unit (communication unit)
- **C** EVA — H. Tracking and Data Relay Satellite System
- **M** ETI — I. Air Force Base
- **K** SRB — J. Kennedy Space Center
- **F** HALS — K. Solid Rocket Booster
- **N** ELV — L. Intravehicular Activity (work done inside crew cabin)
- **J** KSC — M. Extraterrestrial Intelligence (aliens)
- **E** STS — N. Expendable Launch Vehicle (a rocket used only once)
- **A** PLSS — O. Mission Control Center

Make up five "space sentences" using at least one space acronym in each.

1. *(Answers will vary.)*

Space Matching

Match the words on the left with the definitions on the right.

- **J** EMU — A. Takes off like a rocket and lands like a plane
- **G** Goddard — B. Protects the Shuttle from frictional heat
- **L** Sputnik — C. A scientific work area in the cargo bay of the Shuttle
- **K** Shepard — D. Anything that orbits another object
- **M** Glenn — E. A kind of city in space
- **O** Armstrong — F. Feeling of weightlessness
- **A** Space Shuttle — G. He launched the first liquid-fueled rocket
- **B** silicon tiles — H. The equipment on the Space Shuttle
- **F** microgravity — I. A United States person trained for space travel
- **Q** galley — J. Extravehicular Mobility Unit
- **P** PLSS — K. The first U.S. astronaut to orbit the Earth
- **N** cosmonaut — L. The first man-made satellite to orbit the Earth
- **I** astronaut — M. He spent only 15 minutes in space the first time
- **H** payload — N. A Soviet trained for space travel
- **R** crew cabin — O. The first man to walk on the Moon
- **C** Spacelab — P. Contains enough oxygen for seven hours
- **D** satellite — Q. Another name for kitchen aboard the Shuttle
- **E** space station — R. 71.5 cubic meters (2,525 cubic feet) on two floors

SPACE EXPLORATION

Read the clues below and give the name of what it describes. Circle the names in the word search. They may go in any direction.

CLUES
1. The first man on the moon. *Armstrong*
2. A space traveler from the United States. *astronaut*
3. The kitchen aboard the shuttle. *galley*
4. A large storage area for payloads. *cargo bay*
5. The country that made "rockets" in 1232. *China*
6. A space traveler from Russia. *cosmonaut*
7. The first man to orbit the Earth. *Gagarin*
8. A scientist who launched a rocket 41 feet. *Goddard*
9. A cosmonaut that was a dog. *Laika*
10. So little gravity that objects float in the air. *microgravity*
11. A place in spacecraft for dirty laundry and trash. *plastic bag*
12. The country that put the first man into space. *Russia*
13. A reusable spacecraft. *shuttle*
14. A laboratory that fits the cargo bay of the space shuttle. *spacelab*
15. A kind of city that orbits the Earth. *space station*
16. The first rocket to orbit the Earth. *Sputnik*
17. The first woman in space. *Tereshkova*
18. The country that put the first man on the Moon. *United States*

© 1996 Kelley Wingate Publications 128 CD-3728

Answer Key

TRUE OR FALSE?

Read each statement carefully. If the statement is true, put a T on the blank. If the statement is false, put a F on the blank.

T 1. The Space Shuttle can be reused for other missions.
F 2. The external fuel tank is dropped in space where it disappears.
F 3. The external tank carries liquid oxygen and liquid nitrogen.
T 4. The top floor of the crew cabin contains the flight deck.
T 5. Crew members may sleep in a bunk or in a bag on the wall.
F 6. The Spacelab can be found in the cockpit.
F 7. Spacelab was developed by five countries.
F 8. The first space station was called Mir.
F 9. The record for staying in space is 493 days.
T 10. All external fuel tanks are dropped from the orbiter before it goes into orbit around the Earth.
T 11. In space, your waist and feet get smaller.
T 12. Microgravity can give you the sniffles and make you sick or dizzy.
T 13. Over 100 foods are stored in the galley.
F 14. The galley has a refrigerator, but no freezer.
T 15. Crumbs or spills may get into the equipment and damage it.
T 16. The EMU liner is filled with tubes of water to keep the astronaut cool.
T 17. Silicon tiles protect the Space Shuttle from burning up on re-entry.

SPACE EXPLORATION CROSSWORD

Across: GODDARD, CARGOBAY, COSMONAUT, SHUTTLE, SPUTNIK, GALLEY, SPACESTATION
Down: GAGARIN, ASTRONAUT, ARMSTRONG, EMU, SPACELAB, MICROGRAVITY, SPACELAB

ACROSS
1. 1st person to successfully launch a rocket.
3. The part of a shuttle used to carry the payload
5. Spacecraft that can be used again and again
7. Space traveler trained in Russia
9. First rocket to orbit the Earth
11. The kitchen of a spacecraft.
12. A satellite "city" that orbits the Earth.

DOWN
1. Last name of the first man in space
2. A space traveler trained in the United States.
4. First man to walk on the Moon
6. Extravehicular Mobility Unit
8. The feeling of weightlessness
9. An international project used for space experiments
10. A place to store trash and dirty laundry.

WHO AM I?

Read each clue and decide who or what is the answer. Write your answer on the line. Answers may be used more than once!

1. The spacecraft that can be reused. — shuttle
2. The person trained for space travel. — astronaut/cosmonaut
3. The only part of the Shuttle that cannot be reused. — external fuel tank
4. We keep the Shuttle from burning up on re-entry. — silicon tiles
5. Two floors where the crew works and lives. — crew cabin
6. We conduct all experiments during the mission. — mission specialists
7. The laboratory kept in the cargo bay. — Space lab
8. A "hotel" or city that orbits Earth. — space station
9. We hold fuel for lift-off and can be reused. — solid rocket boosters
10. I make you feel weightless. — microgravity
11. The place where food is prepared. — galley
12. A backpack that holds seven hours of oxygen. — PLSS
13. I hold all garbage and soiled clothes so germs don't spread. — plastic bags
14. I am found behind the crew cabin and carry most of the payload, such as the Spacelab. — cargo bay

ACROSTIC

Read the clues and figure out the name of the astronomer. Write his name in the spaces provided. (Clue number one goes first, two second, and so on.)

SPACESUIT
PAYLOAD
GODDARD
MICROGRAVITY
SPACELAB
COSMONAUT
DEHYDRATED
EMU
ARMSTRONG
SPUTNIK
GALLEY
EXTERNALTANK

1. Clothing, gloves, and helmet worn to protect the space crew.
2. Equipment carried aboard the Space Shuttle.
3. The man who launched the first liquid propelled rocket 41 feet in the air.
4. A feeling of weightlessness.
5. A laboratory built by ten countries and carried in the Space Shuttle.
6. A Soviet astronaut.
7. Food that must have the water replaced before it can be eaten.
8. A spacesuit that is worn for work outside the crew cabin.
9. The first man to walk on the surface of the Moon.
10. The first rocket to orbit the Earth.
11. The kitchen in the Space Shuttle.
12. The only part of the Space Shuttle that must be replaced after it is used.

© 1996 Kelley Wingate Publications   129   CD-3728

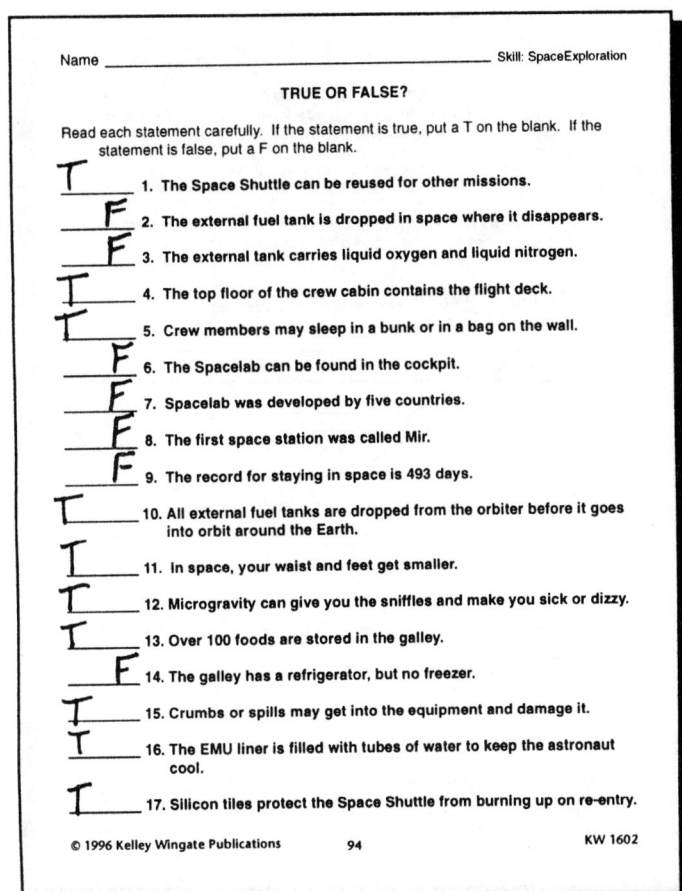

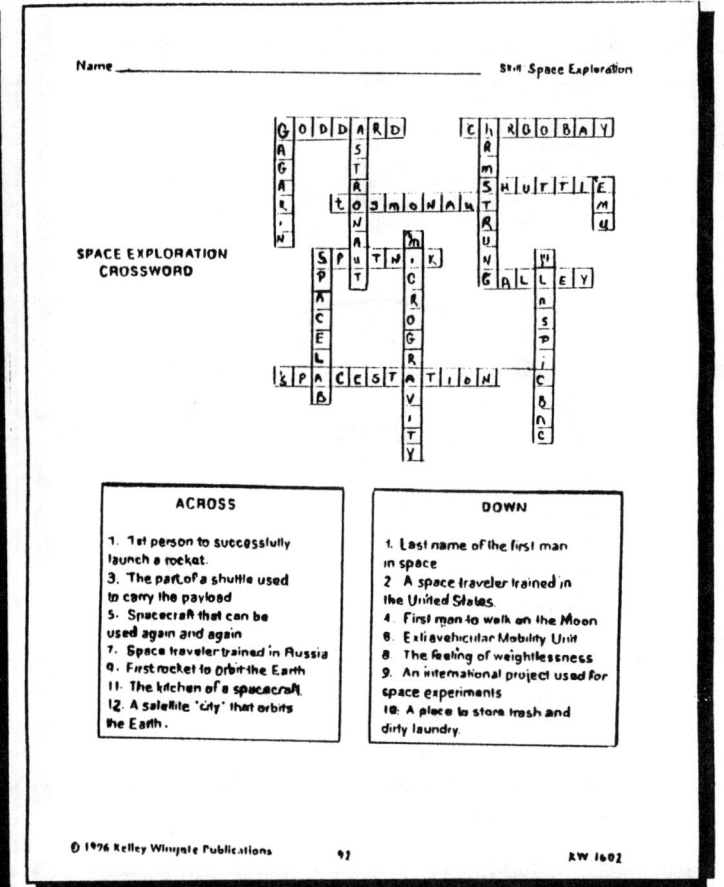

Answer Key

Crew Patch

Every mission that is taken in space has a crew patch designed to represent it. The patch gives the name of the mission, dates of take-off and re-entry, and sometimes the names of the crew members. As a well known artist that will be one of the crew members on the next mission, you have been selected to design the crew patch. You may pick the planet, the members of your crew, and the mission name. Design your patch then draw and color it below.

Answers will vary.

Alien: Friend or Foe?

You have been sent from Earth to visit one of the planets in our solar system. Your mission is to search for life on this planet. After traveling through space for months or years, you finally land. As you depart your spaceship you hear strange sounds nearby. Cautiously you move across the surface of the planet. Suddenly something taps your shoulder. You turn and find yourself face to face with an alien.

Write a brief paragraph describing the alien you are facing. Be sure to include things an alien would need to survive on the planet you chose. Then, draw a picture of your alien on the back of this paper.

Planet landed on: _____

Answers will vary.

How Much Do You Remember?

Finish the statement or answer the question with a word or short answer.

1. A cloud of dust and gas in space is called **nebula**.
2. What is the name of the largest star? **Antares**
3. A star that pulls in its own light is called **a black hole**.
4. What did Galileo locate on the Sun? **sunspots**
5. What is the age of our Sun? **4.6 billion years**
6. A group of stars that make a picture is called **constellation**.
7. Food that is preserved by radiation is called **irradiated**.
8. The tubes in the **EMU** helps keep you cool.
9. **Lift-Off** takes place when a spacecraft is launched.
10. **Van Maanen's** is the smallest known star.
11. Which astronomer wrote the Almagest? **Ptolemy**
12. 186,000 miles per second is **the speed of light**.
13. The planet **Saturn** has the most moons.
14. A **eclipse** occurs when the Moon comes directly between the Sun and Earth.
15. What is the distance to the Sun? **150 million km / 93 million miles**
16. **Venus** is the planet closest to the Earth.
17. Which planet is known as Earth's sister planet? **Venus**
18. A **comet** is like a dirty ball of ice.

19. What comes near the Earth every 76 years? **Halley's Comet**
20. **North Star / Pole Star** is another name for Polaris.
21. About how many food items are carried on the Space Shuttle? **over 100**
22. **Microgravity** makes you feel weightless.
23. The astronomer **Eratosthenes** measured the circumference of the Earth.
24. **Milky Way** is the name of our galaxy.
25. A large group of "shooting stars" is really a **meteor shower**.
26. A gush of fire that rises above the Sun then arches back down is called a **prominence**.
27. The planet **Earth** rotates every 24 hours.
28. William Herschel discovered which planet? **Uranus**
29. This planet takes 248 years to revolve around the Sun. **Pluto**
30. Pieces of rock and metal that orbit the Sun are **asteroids**.
31. **Pluto** is the planet usually furthest from the Sun.
32. The Caloris Basin can be found on **Mercury**.
33. Triton is a moon of **Neptune**.
34. A **meteorite** is a meteor that reaches the Earth.
35. The Great Red Spot is found on the planet **Jupiter**.
36. The **space shuttle** is a reusable spacecraft.
37. The coolest stars are **red** in color.

Answer Key

Name _____ Skill: Recall

38. The **moon** is a satellite that orbits the Earth.
39. **Ptolemy** named 48 of the constellations.
40. The astronomer known as the Father of Systematic Astronomy was **Hipparchus**
41. Mercury is **57.8 million** kilometers from the Sun.
42. A new star is composed mainly of **hydrogen** gas.
43. Galileo found that **Venus** had phases like the Moon, proving the Sun is the center of the solar system.
44. **Mars** is also called "The Red Planet".
45. Most stars are the same size as **our sun**.
46. **Copernicus** introduced heliocentric theory.
47. The Great Red Spot is a **storm** that is over 300 years old.
48. **Aristotle** spent twenty years studying under Plato.
49. Who turned a Dutch toy into a telescope used to observe the stars? **Galileo**
50. The idea that the Sun and all planets revolve around the Earth is called **geocentric** theory.
51. The **moon** is the only natural satellite of Earth.
52. Neptune was discovered in what year? **1846**
53. **Pluto** is the smallest and coldest planet.
54. The astronomer **Eudoxus** developed the geocentric system.

Name _____ Skill: Recall

56. **Hipparchus** discovered a nova and developed accurate instruments for measuring movement of stars and planets.
57. **Heliocentric** theory states that the planets revolve around the Sun.
58. **Methane** gas gives Neptune its blue color.
59. **Brahe** has two wind systems blowing in opposite directions.
60. The moon called **Charon** is almost as large as the planet it orbits.
61. The hottest stars are **blue** in color.
62. **Kepler** developed three laws of planetary motion.
63. The fifth planet from the Sun is **Jupiter**.
64. Our Sun is a **middle-sized** star.
65. **Saturn** has more satellites than any other planet.
66. Newton explained the theories of **gravity** and centrifugal force.
67. Name the four main parts of the Space Shuttle.
   1. **orbiter**
   2. **rocket boosters**
   3. **external fuel tank**
   4. **three main engines**
68. Miranda, Ariel, Umbriel, Titania, and Oberon are the five largest moons of **Uranus**

Name _____ Skill: Recall

69. The highest point on Earth is **Mount Everest**.
70. Jupiter has **3** rings.
71. **Solar winds** push the tail of the comet away from the Sun.
72. The star nearest the Earth is **the sun**.
73. Ursa Major is a constellation with **seven** stars.
74. The Asteroid Belt orbits around the Sun between the planets Jupiter and **mars**.
75. Solar flares give off **radiation** that can affect radio waves on Earth.
76. The **coma** is a cloud of gas that surrounds the nucleus of a comet.
77. Sunspots have an **eleven (11)** year cycle.
78. There is a total of **88** different constellations.
79. **Meteoroids** are bits of dust and rock floating in space, probably left by a passing comet.
80. **China** had rocket arrows in AD 1232.
81. The Big Dipper or Big Bear is also called **Ursa Major**.
82. An old star that suddenly expands and grows very bright is called a **nova**.
83. Asteroids hit the Earth during the **ice age**, but not since that time.
84. A star usually lives for about **10 billion** years.
85. The tail of Halley's comet is about as long as the distance from the Earth to **to the sun**.

Certificate of Completion

_____

receives this award for outstanding performance in

_____

Congratulations!

_____ _____
signed   date

Solar System Award

_____

receives this award for

_____

You are terrific!

_____ _____
signed   date

©1996 Kelley Wingate Publications

absorb	alien	Antares	asteroid
astronaut	astronomer	atmosphere	axis
black hole	boosters	cargo	celestial
circumference	comet	concentric	constellation

cosmonaut	crater	crew	cycle
dehydrated	diameter	discovery	distance
dwarf	Earth	eclipse	elliptical
exploration	external	flare	formula

galaxy	Galileo	galley	gas planet
geocentric	heliocentric	helium	hydrogen
Jupiter	laboratory	latitude	launch
light year	longitude	lunar	magnetic

Mars	Mercury	meteor	meteorite
meteoroid	methane	microgravity	Milky Way
mission	nebula	Neptune	nitrogen
nova	observe	orbit	payload

planet	pole	Polaris	Pluto
probe	reflect	radiate	prominence
revolution	rotation	rotate	revolve
satellite	shuttle	scientist	Saturn

solar	space craft	spacelab	sphere
Sputnick	sunspot	supernova	telescope
temperature	theory	transportation	universe
Uranus	Ursa Major	Venus	weightless